Mohammad H. Keshavarz, Thomas M. Klapötke
The Properties of Energetic Materials
De Gruyter Studium

Also of Interest

High Explosives, Propellants, Pyrotechnics
Koch; 2021
ISBN 978-3-11-066052-4, e-ISBN 978-3-11-066056-2

Chemistry of High Energy Materials
Klapoetke; 2019
ISBN 978-3-11-062438-0, e-ISBN 978-3-11-062457-1

Energetic Compounds
Methods for Prediction of their Performance
Keshavarz, Klapoetke; 2020
ISBN 978-3-11-067764-5, e-ISBN 978-3-11-067765-2

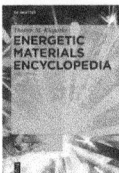

Energetic Materials Encyclopedia
2nd Edition. 3 Volumes
Klapötke; 2021
ISBN 978-3-11-062488-5, e-ISBN 978-3-11-062681-0

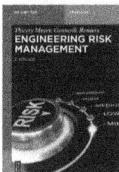

Engineering Risk Management
2nd Edition
Meyer, Reniers; 2016
ISBN 978-3-11-041803-3, e-ISBN 978-3-11-041804-0

Mohammad H. Keshavarz, Thomas M. Klapötke

The Properties of Energetic Materials

—

Sensitivity, Physical and Thermodynamic Properties

2nd edition

DE GRUYTER

Authors

Prof. Dr. Mohammad H. Keshavarz
Malek-ashtar University of Technology
Department of Chemistry
83145 115 Shahin-shahr
Iran
keshavarz7@gmail.com

Prof. Dr. Thomas M. Klapötke
Ludwig-Maximilians-Universität
Department of Chemistry/
Energetic Materials Research
Butenandstr. 5-13
81377 Munich
Germany
tmk@cup.uni-muenchen.de

ISBN 978-3-11-074012-7
e-ISBN (PDF) 978-3-11-074015-8
e-ISBN (EPUB) 978-3-11-074024-0

Library of Congress Control Number: 2021941395

Bibliographic information published by the Deutsche Nationalbibliothek
The Deutsche Nationalbibliothek lists this publication in the Deutsche Nationalbibliografie;
detailed bibliographic data are available on the Internet at http://dnb.dnb.de.

© 2021 Walter de Gruyter GmbH, Berlin/Boston
Cover image: Jag_cz / iStock / Getty Images Plus
Typesetting: VTeX UAB, Lithuania
Printing and binding: CPI books GmbH, Leck

www.degruyter.com

Preface

For a chemist who is concerned with the synthesis of new energetic compounds, it is essential to be able to assess physical and thermodynamic properties, as well as the sensitivity of possible new energetic compounds before synthesis is attempted. Various approaches have been developed to predict important aspects of the physical and thermodynamic properties of energetic materials including (but not exclusively): crystal density, heat of formation, melting point, enthalpy of fusion and enthalpy of sublimation of an organic energetic compound. Since an organic energetic material consists of metastable molecules capable of undergoing very rapid and highly exothermic reactions, many methods have been developed to estimate the sensitivity of an energetic compound with respect to detonation-causing external stimuli such as heat, friction, impact, shock, and electrostatic discharge. This book introduces these methods and demonstrates those methods which can be easily applied.

https://doi.org/10.1515/9783110740158-201

Preface to the second edition

Everything said in the preface to the first edition still holds and essentially does not need any addition or correction. In this revised second edition, we have updated the manuscript and added some recent aspects of energetic materials:

1. Some errors which unfortunately occurred in the first edition have been corrected and the references have been updated where appropriate.
2. Recent works have been reviewed and discussed in each chapter. Moreover, new sections have been inserted including:
   (a) Chapter 1 – The use of group additivity methods for prediction of crystal density of energetic neutral and ionic liquids or salts
   (b) Chapter 2 – The condensed phase heat of formation of energetic ionic liquids and salts
   (c) Chapter 3 – Melting points of ionic liquids
   (d) Chapter 4 – Group additivity method for prediction of enthalpy and entropy of fusion
   (e) Chapter 5 – Group additivity method for prediction of the heat of sublimation
   (f) Chapter 6 – Impact sensitivity of quaternary ammonium-based energetic ionic liquids or salts
   (g) Chapter 7 – Simple prediction of electrostatic spark sensitivity based on the new ESZ KTTV instrument
   (h) Chapter 8 – Critical diameter of solid pure and composite high explosives
   (i) Chapter 9 – Friction sensitivity of quaternary ammonium-based energetic ionic liquids
   (j) Chapter 10 – Thermal stability of selected classes of energetic ionic liquids and salts
   (k) Chapter 11 – A general correlation between electric spark sensitivity and impact sensitivity of nitroaromatics and nitramines as well as the relationship between shock sensitivity of nitramine energetic compounds based on small-scale gap test and their electric spark sensitivity

Mohammad Hossein Keshavarz
Thomas M. Klapötke

https://doi.org/10.1515/9783110740158-202

About the authors

Mohammad Hossein Keshavarz

Mohammad Hossein Keshavarz born in 1966, studied chemistry at Shiraz University and received his BSc in 1988. He also received a MSc and PhD at Shiraz University in 1991 and 1995. From 1997 until 2008, he was Assistant Professor, Associate Professor and Professor of Physical Chemistry at the University of Malek-ashtar in Shahin-shahr of Iran. Since 1997 he is Lecturer and researcher at the Malek-ashtar University of Technology, Iran. He is the editor of two research journals in the Persian language. Keshavarz has published over 400 scientific papers in international peer reviewed journals, 5 book chapters and eight books in the field of the assessment of energetic materials (four books in Persian language and four books in English language.

Thomas M. Klapötke

Thomas M. Klapötke received his PhD in 1986 (TU Berlin), post-doc in Fredericton, New Brunswick, habilitation in 1990 (TU Berlin). From 1995 until 1997 Klapötke was Ramsay Professor of Chemistry at the University of Glasgow in Scotland. Since 1997 he has held the Chair of Inorganic Chemistry at LMU Munich. In 2009 Klapötke was appointed a Visiting Professor at CECD, University of Maryland and in 2014 he was appointed a Adjunct Professor at the University of Rhode Island. Klapötke is a Fellow of the RSC (C. Sci., C. Chem. F. R. S. C.), a member of the ACS and the Fluorine Division of the ACS, a member of the GDCh, and a Life Member of both the IPS and the National Defense Industrial Association. Most of Klapötke's scientific collaborations are between LMU and ARL (US Army Research Laboratory) in Aberdeen, MD and ARDEC (Armament Research Development and Engineering Center) in Picatinny, NJ. Klapötke also collaborates with ERDC in Champaign, IL. And Prof Ang, How-Ghee (NTU, Singapore). He is the executive editor of the Journal of Engineering Science and Military Technologies, the Subject Editor in the area of explosives synthesis of the Central European Journal of Energetic Materials and an editorial board member of Propellants, Explosives and Pyrotechnics (PEP), Journal of Energetic Materials, the Chinese Journal of Explosives and Propellants and the International Journal of Energetic Materials and Chemical Propulsion (IJEMCP). Klapötke has published over 850 papers, 33 book chapters and 17 books.

https://doi.org/10.1515/9783110740158-203

Contents

1 Crystal density

Organic compounds containing energetic groups such as nitro, nitramine, and nitrate ester functional groups have wide applications in military and civilian applications as propellants, explosives, and pyrotechnics because they can release their stored chemical energy upon external stimuli such as heat, impact, shock, friction, and electrostatic discharge [1–7]. Ionic molecular energetic materials containing high nitrogen content are attractive for scientists and industries. They may be used as energetic compounds because they have high density, positive heats of formation, and thermal stability [8]. They frequently consist of high nitrogen content cations such as substituted imidazole, triazole, and tetrazole derivatives and bulky anions containing energetic groups, e. g. $-NO_2$, $-N_3$, and $-CN$. They can be considered as eco-friendly, low-melting, and thermally stable ionic compounds [2]. Considerable efforts have been done in recent years to introduce new organic and ionic molecular energetic materials with high density because their higher density is always desirable for packing more energy per unit volume.

The crystal density and condensed phase heat of formation of an energetic compound are two important physicothermal properties, and are essential values in order to be able predict the detonation performance using a thermodynamic equilibrium code such as CHEETAH [9], or through empirical methods [1, 10–17]. The performance characteristics of energetic compounds are proportional to their densities, e. g. the Chapman–Jouguet pressure is proportional to the square of the initial density [1, 10]. Thus, it is essential to use suitable methods such as gas pycnometry or low-temperature single crystal x-ray diffraction to determine the crystal density of an energetic compound. New molecules which are candidates for possible use as energetic materials can be synthesized, characterized, and formulated by reliable predictive methods, and theoretical molecular design may be used to develop new

https://doi.org/10.1515/9783110740158-001

energetic materials before synthesis is attempted. The synthesis of molecules with significantly increased energy in comparison with current materials, as well as the synthesis of very insensitive materials which have reasonable energies, have been two important goals for scientists in recent years. Since it is essential to have reliable methods to predict the density of energetic compounds, different approaches have been developed to assess the crystal density of an energetic compound at 25 °C.

Attempts have been made to predict the crystal densities of proposed new energetic compounds with satisfactory accuracy. A predicted crystal density that differs by less than $0.03\,g/cm^3$ from the experimentally obtained value should be defined as "excellent". A value that deviates between 0.03 and $0.05\,g/cm^3$ from the experimental value is still "informative" [18]. Quantum-mechanically determined molecular volumes [19–22], group additivity [23–25], empirical methods [26–29], quantitative structure–property relationships (QSPR) based on complex descriptors [30, 31] and molecular dynamics (MD) [32] are usual different approaches which have developed to predict the crystal densities of different types of $C_aH_bN_cO_d$ energetic compounds. The group additivity method is a simple approach because it requires only a set of atoms and group volumes that can be summed to obtain an estimate of the effective molecular volume using a simple computer code [24]. Tedious investigations have been undertaken over the last 30 years or so to expand the list of atom and functional group volumes [24, 25]. Although group additivity methods are simple to use with low cost, the predicted value which is obtained may show very large deviation from the experimentally obtained value for some energetic compounds. Moreover, such methods can only be used for those energetic compounds for which the values of all groups contained in the compound have been specified. Quantum mechanical and empirical methods (or QSPR) based on the structures of energetic compounds are more reliable approaches to estimate the density of an energetic compound. The QSPR methods are based on complex descriptors which develop a mathematical relationship connecting a macroscopic property of a series of compounds to microscopic descriptors derived from their molecular structures using an experimental data set. They require computer codes and expert users, as well as complex descriptors. The descriptors used in QSPR models can be empirical, or computed on the basis of the molecular structure. Various statistical tools including multilinear regression (MLR), nonlinear regression (NLR), partial least squares (PLS), artificial neural network (ANN), genetic algorithm (GA) or support vector machine (SVM) are frequently used to derive the mathematical equations (or algorithms) linking the property and descriptors [33, 34]. MD is a computer simulation of the physical movements of atoms and molecules in the context of N-body simulation. Due to the higher reliability of quantum mechanical and empirical methods (or QSPR) based on molecular/ionic structures, these approaches have been developed in recent years for neutral and ionic liquid energetic compounds, which are described in this chapter. Some efforts have been made to assess the detonation performance of newly designed explosives and ionic molecular energetic materials with high detonation performance in recent years [13, 15, 35–46]. Since high reliability is an important parameter in selecting predictive methods for different classes of energetic compounds, several of the best available methods are introduced and described in this chapter.

1.1 Group additivity method

The group additivity method sums the volume of atoms, molecular fragments, and functional groups to estimate the molecular volume of an organic energetic compound. It cannot explain the effect of parameters including the molecular conformation, isomerism, and packing efficiency in the crystals on density. Moreover, it provides the same results for organic explosives with different isomers and conformations. Thus, the reliability of group additivity methods is low as compared to the other common methods. Ammon [47] has introduced the latest group additivity method by including a larger database of groups, i. e. 96 different groups and atoms from more than 26,000 crystals. Ye and Shreeve [48] introduced a suitable group additivity method that only involves 38 atom/group parameters and three corrections. This approach considers the quantitative impact of strong hydrogen bonding on the densities of energetic materials. Table 1.1 shows volume parameters for atoms, groups, and fragments. The sum of the contributions of these volume parameters can be used to estimate density at room temperature as

$$\rho = \frac{Mw}{0.6022V}, \tag{1.1}$$

where Mw is the molecular weight of the desired explosive in g/mol and V is the total volume in $Å^3$, respectively.

Example 1.1. N,N-bis(2-fluoro-2,2-dinitroethyl)nitramide has the following structure:

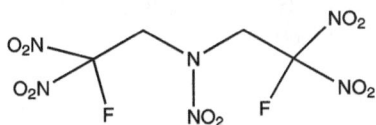

Molecular Weight: 334.11

The use of equation (1.1) and Table 1.1 gives:

$$V = V(NO_2) \times 5 + V(F) \times 2 + V(CH_2, \text{acyclic}) \times 2 + V(C, \text{acyclic}) \times 2 + V(N)$$
$$= 36 \times 5 + 12.5 \times 2 + 24 \times 2 + (24 - 10) \times 2 + 10 = 291 \, Å^3$$
$$\rho = \frac{Mw}{0.6022V} = \frac{334.10}{0.6022 \times 291} = 1.907 \, \text{g/cm}^3.$$

The measured X-ray density of this compound is 1.917 g cm^{-3} [48].

Ye and Shreeve [48] considered corrections of sp^3 C or sp^3 N in two or more fused rings as well as sp^3 C in two or more caged rings and sp^2 C in two or more rings, which are important for the design of high energy density materials (HEDMs).

Table 1.1: Volume parameters for atoms, groups, and fragments based on the method of Ye and Shreeve [48].

Species	Volume (Å^3)	Species	Volume (Å^3)
Neutral			
Imidazole	84	1,2,4-triazole	79
Tetrazole	75	s-triazine	90
1,2,4,5-tetrazine	87	pyrimidine	100
Cubane	135	furazan	77
Benzene	110	pyridine	105
Groups			
H (bonded to N)	7	H (bonded to C)	5
CH_3	30	CH_2 (acyclic)	24
CH_2 (three- or four-membered ring)	22.5	CH_2 (five- or six-membered ring)	22
CH_2 (eight-membered ring)	21	CH (in isoWurtzitane)[a]	13.5
–C=C–	26.5	–C=N–	25
CN	30	–N=N–	26
NH_2	20	NO_2	36
NH	15	N_3	41
N (in tetraazapentalene)	9.5	N (in other cases)	10
C=O (not in a ring)	25	C=O (in a ring)	22
COOH	41	OH	15
O (in ether or –O–NO2)	11.5	O (in other cases)	10
F	12.5	CF_2	37.5
NF_2	37	SF_5	82
Corrections			
strong hydrogen bonds for each NH_2 or NH	−8		
each sp^3 C or sp^3 N in two or more rings	−1		
each sp^2 C in two or three rings	−2		

[a]Volumes of other CH moieties were derived from respective CH_2: $V(CH) = V(CH_2) - 5/\text{Å}^3$; while $V(C) = V(CH_2) - 10/\text{Å}^3$. [b]Except in tetrazole where sp^2 C does not need be corrected.

Example 1.2. Consider the following three HEDMs:

Molecular Weight: 232.11

(a) GEMZAZ, aka DINGU

Molecular Weight: 416.21

(b) JAJBEB

Molecular Weight: 388.21

(c) PUTCEM, aka z-TACOT

Necessary corrections for sp^3 C, sp^3 N and sp^2 C in these compounds are marked with an asterisk. The use of equation (1.1) and Table 1.1 for three compounds gives:

(a) Volume of each asterisk CH needs to be corrected by $-1\,\text{Å}^3$ in GEMZAZ.

$$V = V(NO_2) \times 2 + V(C=O, \text{in ring}) \times 2 + V(NH) \times 2 + V(N) \times 2 + V(CH) \times 2 - 1 \times 2$$

$$= 36 \times 2 + 22 \times 2 + 15 \times 2 + 10 \times 2 + (22 - 5) \times 2 - 2 = 198\,\text{Å}^3$$

$$\rho = \frac{Mw}{0.6022V} = \frac{232.11}{0.6022 \times 198} = 1.947\,\text{g/cm}^3.$$

The measured X-ray density of GEMZAZ is 1.99 g/cm^3 [48].

(b) There are eight asterisk −CH groups in three rings of JAJBEB. Thus, the volume of each −CH must be corrected by $-1\,\text{Å}^3$.

$$V = V(NO_2) \times 2 + V(C=O, \text{in ring}) \times 2 + V(NH) \times 2 + V(N) \times 2 + V(CH) \times 2 - 1 \times 2$$

$$= 36 \times 2 + 22 \times 2 + 15 \times 2 + 10 \times 2 + (22 - 5) \times 2 - 2 = 380\,\text{Å}^3$$

$$\rho = \frac{Mw}{0.6022V} = \frac{416.21}{0.6022 \times 380} = 1.819\,\text{g/cm}^3.$$

The measured X-ray density of JAJBEB is 1.828 g/cm^3 [48].

(c) There are four sp2 C in two rings of PUTCEM. Each requires correction of $-2\,\text{Å}^3$ in volume.

$$V = V(\text{benzene}) \times 2 - V(H) \times 8 + V(NO_2) \times 4 + V(N, \textit{in tetraazapentalene}) \times 4 - 2 \times 4$$

$$= 110 \times 2 - 5 \times 8 + 36 \times 4 + 9.5 \times 4 - 2 \times 4 = 354\,\text{Å}^3$$

$$\rho = \frac{Mw}{0.6022V} = \frac{388.21}{0.6022 \times 354} = 1.821\,\text{g/cm}^3.$$

The reported X-ray density of PUTCEM is 1.830 g/cm^3 [48].

Ye and Shreeve [48] corrected the volume of each NH$_2$ or NH group by $-8\,\text{Å}^3$ for strong bonding in three categories: (i) both carbons vicinal to the NH$_2$ or NH group have nitro groups or one C−NO$_2$ and one N-oxide (N-O). Due to electronic and/or steric effects, this condition cannot be applied if the NH$_2$ group was substituted by an alkyl, phenyl, or other electrondonating groups, e. g. *N*-methyl-2,4,6-trinitrobenzenamine (JUPROB) or *N*-methyl-2,6-dinitro-4-(trifluoromethyl) benzenamine (FMANIL).

Example 1.3. The two −NH$_2$ groups of N^1-isopropyl-2,4,6-trinitrobenzene-1,3,5-triamine with the following molecular structure require correction for hydrogen bonding but there is no need to consider correction of NH group because it is attached to a isopropyl group (electron donating) on −NH.

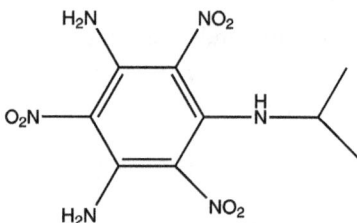

Molecular Weight: 300.23

The use of equation (1.1) and Table 1.1 provides:

$$V = V(\text{benzene}) - V(\text{H}) \times 6 + V(\text{NO}_2) \times 3 + V(\text{NH}_2) \times 2 + V(\text{NH}) + V(\text{isopropyl})$$
$$- V(\text{hydrogen bonding}) \times 2$$

$$= 110 - 5 \times 6 + 36 \times 3 + 20 \times 2 + 15 + (30 \times 2 + 24 - 5) - 8 \times 2 = 306\,\text{Å}^3$$

$$\rho = \frac{Mw}{0.6022V} = \frac{300.23}{0.6022 \times 306} = 1.629\,\text{g/cm}^3.$$

The reported value of the X-ray density of this compound is 1.604 g/cm^3 [48].

(ii) There is no need to correct the volume of the amino group remains at 20 Å^3 in Table 1.1 when only one vicinal carbon bears a nitro group. Meanwhile, if the molecule has $C2$ symmetry, the volume correction of NH$_2$ groups by −8 Å^3 should be considered. At least two nitro and two NH$_2$ groups exist in the molecule and in the vicinal position here. Three examples for this situation are 1,1-diamino-2,2-dinitroethene (FOX-7), 2,4,6-triamino-3,5-dinitropyridine (TIBMUM), and 2,6-diamino-3,5-dinitropyrimidine (CIWMAW01).

Example 1.4. FOX-7 has the following molecular structure:

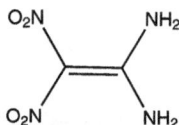

Molecular Weight: 148.08

The use of equation (1.1) and Table 1.1 provides:

$$V = V(\text{-C=C-}) + V(\text{NO}_2) \times 2 + V(\text{NH}_2) \times 2 - V(\text{hydrogen bonding})$$

$$= 26.5 + 36 \times 2 + 20 \times 2 - 8 = 130.5\,\text{Å}^3$$

$$\rho = \frac{Mw}{0.6022V} = \frac{148.08}{0.6022 \times 130.5} = 1.884\,\text{g/cm}^3.$$

The measured value of density of this compound is 1.883 g/cm^3 [48].

(iii) The volume of the molecule also needs a correction of −8 Å if some heterocycles (triazole, pyrazole, etc.) include an −NHNO$_2$ group and the heterocycle also has an acidic N–H on the ring, e. g. 5-nitramino-1,2,4-triazole (NRTZ) and 3-nitramino-4,5-dinitro-pyrazole.

Example 1.5. 5-Nitramino-1,2,4-triazole (NRTZ) with the following molecular structure follows this condition.

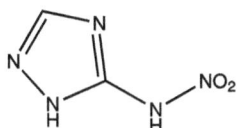

Molecular Weight: 129.08

The use of equation (1.1) and Table 1.1 provides:

$$V = V(1,2,4\text{-triazole}) - V(\text{H}) + V(\text{NO}_2) + V(\text{NH}) - V(\text{hydrogen bonding})$$

$$= 79 - 5 + 36 + 15 - 8 = 117\,\text{Å}^3$$

$$\rho = \frac{Mw}{0.6022V} = \frac{129.08}{0.6022 \times 117} = 1.832\,\text{g/cm}^3.$$

The reported value of density of NRTZ is 1.83 g/cm^3 [48].

Two further new group additivity models were introduced recently for some specific classes of energetic compounds, which have been illustrated here.

1.1.1 The method of atomic contributions

The method of atomic contributions (MAC) uses the specific molar volume as a sum of volumes of individual atoms constituting the crystal [49]. It can improve the accuracy of the calculations of the molecular crystal density, which contains both explosive and nonexplosive compounds. Thus, the density of molecular crystals is given as a ratio of molar mass to molar volume:

$$\rho = \frac{Mw}{\left(\sum_{A=C,N,O\ldots} \sum_{i=1}^{n_i} B_i \frac{M_i}{\rho_i} + \sum_{j=4}^{n_2} L_j C_j N_j + \sum_{k=1}^{n_3} R_k D_k N_k\right)}, \tag{1.2}$$

where ρ is the density of the compound, Mw is the molecular weight of the compound, B_i is the number of ith atoms in one mole, M_i is the mass of ith atom, C_j is a correction for interactions in cycles, N_j is the number of atoms in a cycle of jth order, L_j is the number of cycles of jth order in one mole, D_k is a correction for atomic functional groups of kth type, N_k is the number of atoms in the group of k type, and R_k is the number of groups of k type in one mole. This method is more complex than the other group additivity methods. Smirnov et al. [49] used this approach for the calculation of the density of several nitramine compounds and energetic compounds containing the oxadiazole ring.

1.1.2 Benzene-derived energetic compounds using atomic volumes

Hofmann [50] reported that the crystal density of a neutral or ionic compound can be calculated by using the average atomic volumes of various elements as well as thermal expansion as follows:

$$\rho = \frac{Mw}{0.01387a + 0.00508b + 0.0118c + 0.01139d} \times 0.00164, \tag{1.3}$$

where a, b, c, and d are the number of carbon, hydrogen, nitrogen, and oxygen atoms; Mw is the molecular weight of the desired explosive. Ghule et al. [51] indicated that equation (1.2) should be revised for those energetic compounds with strong H-bonding or with strong van der Waals or electrostatic interactions. It has been suggested to account for H-bonding between amino and oxygen-containing groups in group additivity methods because they underestimate the density [29, 52, 53]. Ghule et al. [51] found that the existence of two or more $-NH_2$ or $-NH_3^+$ groups in the molecule increases. They considered the contributions of these groups to improve the reliability of equation (1.2) for those benzene-derived energetic compounds containing $-NH_2$ and/or $-NH_3^+$ groups. Thus, they introduced the following equation for estimation of densities of neutral nitrobenzenes, energetic salts, and cocrystals:

$$\rho = \frac{Mw}{0.01387a + 0.00508b + 0.0118c + 0.01139d} \times 0.00175. \tag{1.4}$$

This equation is the simplest approach for the calculation of densities of the mentioned classes of energetic compounds.

Example 1.6. 2-[nitro(2,4,6-trinitrophenyl)amino]ethyl nitrate, 2,4,6-trinitrobenzene-1,3-diamine (DATB) and 2-isopropyl-6-methylpyrimidin-1,3-diium-4-olate picrate have the following structures:

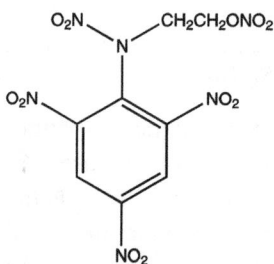

Chemical Formula: $C_8H_6N_6O_{11}$
Molecular Weight: 362.17

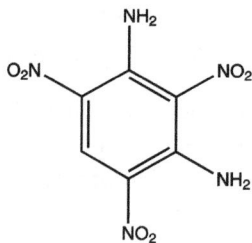

Chemical Formula: $C_6H_5N_5O_6$
Molecular Weight: 243.13

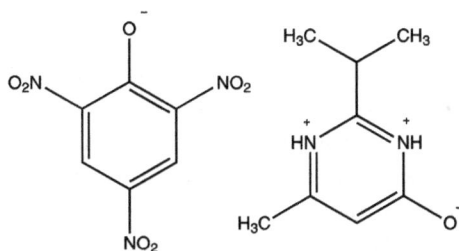

Chemical Formula: $C_{14}H_{15}N_5O_8$
Molecular Weight: 381.30

Equation (1.2) should be used for 2-[nitro(2,4,6-trinitrophenyl)amino]ethyl nitrate and 2-isopropyl-6-methylpyrimidin-1,3-diium-4-olate picrate because there is no $-NH_2$ and/or $-NH_3^+$ groups in their benzene rings:

$$\rho = \frac{Mw}{0.01387a + 0.00508b + 0.0118c + 0.01139d} \times 0.00164$$
$$= \frac{362.17}{0.01387 \times 8 + 0.00508 \times 6 + 0.0118 \times 6 + 0.01139 \times 11} \times 0.00164$$
$$= 1.760 \, \text{g/cm}^3$$

$$\rho = \frac{Mw}{0.01387a + 0.00508b + 0.0118c + 0.01139d} \times 0.00164$$
$$= \frac{381.30}{0.01387 \times 14 + 0.00508 \times 15 + 0.0118 \times 5 + 0.01139 \times 8} \times 0.00164$$
$$= 1.487 \, \text{g/cm}^3$$

The reported crystal densities for 2-[nitro(2,4,6-trinitrophenyl)amino]ethyl nitrate and 2-isopropyl-6-methylpyrimidin-1,3-diium-4-olate picrate are 1.75 [54] and 1.44 g/cm³ [55], respectively.

Equation (1.3) should be used for DATB because there are two $-NH_2$ groups in its benzene ring:

$$\rho = \frac{Mw}{0.01387a + 0.00508b + 0.0118c + 0.01139d} \times 0.00175$$
$$= \frac{243.14}{0.01387 \times 6 + 0.00508 \times 5 + 0.0118 \times 5 + 0.01139 \times 6} \times 0.00164$$
$$= 1.803 \, \text{g/cm}^3$$

The reported crystal density for DATB is 1.83 g/cm³ [23].

1.1.3 The method of group additivity for estimating densities of energetic ionic liquids and salts

Ye and Shreeve [48] introduced a group additivity method for energetic ionic liquids and energetic ionic salts. They applied the method of the closed packed volume of single ions to estimate the density of 59 ionic compounds containing energetic ionic liquids and energetic ionic salts at room temperature. They developed volume parameters, which depend on the packing of the component ions as well as the size and shape of the ions and the ion–ion interactions. This approach can be used only for limited classes of energetic ionic liquids and energetic ionic salts because many volume parameters of groups and fragments were not defined. The existence of energetic groups such as $-NO_2$, $-N_3$, and $-CN$ can increase strong H-bond interactions in a desired energetic ionic liquid or energetic ionic salt. Ye and Shreeve [48] introduced the effective volumes of cations and anions for calculation of molecular volume (V) of room-temperature energetic ionic liquids and energetic ionic salts with general formula MpNq as:

$$V = pV^+ + qV^-,\qquad(1.5)$$

where V^+ and V^- are the effective volumes of cations and anions, respectively. Equation (1.1) can be used to estimate the density. Table 1.2 shows volume parameters of groups and fragments for ionic liquids and salts at room temperature. As seen, the volume parameters for ionic liquids and salts are different for some cations and anions.

Table 1.2: Volume parameters of groups and fragments for ionic liquids and salts at room temperature.

Species	Volume ($Å^3$)	Species	Volume ($Å^3$)
Cations			
1,3-2H-imidazolium (+)[a]	79	1,4-2H-1,2,4-triazolium (+)	73
1,4-2H-tetrazolium (+)	66	N–H-pyridinium (+)[a]	95
guanidinium (+)	69	triaminoguanidinium (+)	105
Me_4N^{+}[a]	113	Me_4P^{+}[a]	133
HMTA–H(+)[b]	155	NH_4^{+}[c]	21
NH_2NH_3 (+)	30	azetidinium (+)	76
Anions			
imidazolate (−)	93	1,2,4-triazolate (−)	87
1-tetrazolate (−)	80	$(NO_2)_3C$ (−)	141
picrate (−)	218	NTO (−)[d]	123
NO_3^-	64	ClO_4^-	82
$N(NO_2)_2^-$	98	N_3^-	60
NO_2^-	55	CN^-	50
CF_3CO_2 (−)	108	CH_3SO_3 (−)	99
TfO^-	129	PF_6^-	107
BF_4^-	73	Tf_2N^{-}[a]	230
Br^-	56	Cl^-	47
Groups			
CH_3[a]	30	C=O (not in a ring)	24
NH_2[e]	20	N	10

[a]For ionic liquids: 1,3-dimethylimidazolium: 154; N-methylpyridinium: 146; Me_4N^+: 136; Me_4P^+: 163; 1,1-dimethylpyrrolidinium: 169; Tf_2N^-: 248; CH_2: 28; CH_3: 35; H 7 $Å^3$. [b]HMTA, hexamethylenetetramine. [c]$V(NH_4^+)$ 15 $Å^3$ for inorganic salts. [d]NTO: 3-nitro-1,2,4-triazol-5-one. [e]For very strong hydrogen bonds, $V(NH_2)$ 12 $Å^3$.

Example 1.7. Consider the following energetic salt:

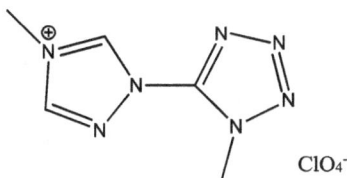

ClO_4^-

The structure of the cation should be considered as follows:

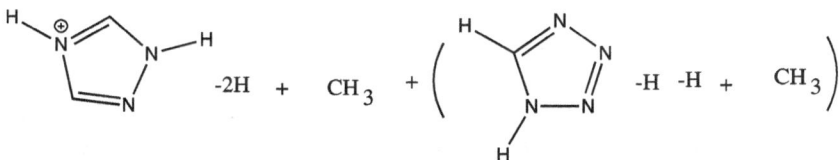

$-2H \quad + \quad CH_3 \quad + \quad \left(\quad \quad -H \;\; -H \;\; + \quad CH_3 \right)$

The use of equations (1.1) and (1.5) as well as Tables 1.1 and 1.2 gives:

$$V^+ = V(1,2,4\text{-triazolium}) - 2 \times V(H, \text{bonded to N}) + V(CH_3) + V(\text{tetrazole})$$
$$- V(H, \text{bonded to C}) - V(H, \text{bonded to N}) + V(CH_3)$$
$$= 73 - 2 \times 7 + 30 + 75 - 5 - 7 + 30 = 182 \,\text{Å}^3$$
$$V^- = V(ClO_4^-) = 82 \,\text{Å}^3$$
$$V = V^+ + V^- = 182 + 82 = 264 \,\text{Å}^3$$
$$\rho = \frac{Mw}{0.6022V} = \frac{265.61}{0.6022 \times 264} = 1.671 \,\text{g/cm}^3.$$

The reported crystal density for this energetic salt is 1.678 g/cm^3 [48].

1.2 Quantum mechanical approach

1.2.1 Quantum mechanical approach for neutral energetic compounds

Quantum mechanical computations require high-speed computers to conduct complicated calculations, and can be used for energetic molecules with simple molecular structures [18, 20–22, 56]. Molecular surface electrostatic potential (MESP) was introduced by Politzer et al. [22, 57] and accounted for the intermolecular interactions in the crystal. It has been widely used in recent years to estimate the density of many classes of $C_aH_bN_cO_d$ energetic compounds. The effectiveness of different MESP-based methods has been advanced to improve the reliability of estimations [18, 20–22, 56, 58–63]. Rice et al. [62] reviewed different quantum mechanical approaches for the prediction of crystal densities of neutral molecular and ionic molecular crystals. Qiu et al.

[21] introduced a very direct approach to estimate the crystal densities of nitramines based on the ratio of the molecular mass and the volume of the isolated gas phase molecule. Rice et al. [20] developed the method of Qiu et al. [21] for primarily nitroaromatics, nitramines, nitrate esters and nitroaliphatics. They presented a method for estimating the densities of neutral and ionic molecular crystals using the quantum-mechanically determined molecular volume of an isolated molecule or formula unit within the crystal as

$$\rho = \frac{M}{V_m},$$ (1.6)

where ρ is the crystal density of energetic compound; M is the molecular mass of the molecule in g/molecule and V_m is the volume inside the 0.001 a. u. isosurface of electron density surrounding the molecule which is calculated using density functional theory (DFT) at the B3LYP/6-31G** level with the Gaussian program package. Politzer et al. [22] improved the procedure presented by Rice et al. [20] by adding corrections for electrostatic interactions, in order to better represent the intermolecular interactions in both neutral and ionic molecular crystals. Rice et al. [62] used a suitable DFT method based on the Politzer et al. method [22] for calculating the density of energetic compounds through the interaction index $v\sigma_{tot}^2$ as

$$\rho = \alpha_1\left(\frac{M}{V_m}\right) + \beta_1(v\sigma_{tot}^2) + \gamma_1,$$ (1.7)

where ρ is in g/cm^3; σ_{tot}^2 and v are the total variance of the electrostatic potential on the 0.001 a. u. molecular surface and the degree of balance between the positive and negative potentials on the molecular surface, respectively. The parameter v calculates the degree of balance between the positive and negative potentials on the molecular surface. The parameter $v\sigma_{Tot}^2$ has a significant contribution to analytical relations of condensed phase properties which depends on intermolecular interactions. For the model of Politzer et al. [22], the values of three parameters α_1, β_1, and γ_1 are 0.9183, 0.0028, and 0.0443, respectively. Meanwhile, the values of these parameters are α_1 = 1.0462, β_1 = 0.0021, and γ_1 = −0.1586 for the model of Rice–Byrd [62]. Politzer et al. [22] optimized the structure and the surface properties at the B3PW91/6-31G(d,p) level whereas Rice–Byrd [62] applied the B3LYP/6-31G** level for calculations. Wang et al. [64] compared the computed densities for thirty-one aliphatic nitrates at room temperature using the density functional theory method (B3LYP) in combination with six basis sets (3-21G, 6-31G, 6-31G*, 6-31G**, 6-311G* and 6-31+G**) and the semiempirical molecular orbital method (PM3). They recommended the B3LYP/6-31G* method because it provided more reliable results to predict the crystalline densities of organic nitrates as compared to those obtained by the QSPR-based method of Keshavarz and Pouretedal [28] as well as MESP-based method Rice–Byrd [62]. Wang et al. [65] used experimental data of 3694 nitro compounds to introduce a molecular morphology descriptor and a hydrogen-bond descriptor as correction items of equation (1.5) to build three new density-functional theory (DFT)-QSPR models. For about 91–93 % of nitro compounds, the percent of deviations of the predicted results are less than 5 % at two levels of B3PW91/6-31G(d,p) and B3LYP/6-31G**. Details of the coefficients α_1, β_1 and γ_1 (which can be found by least square fitting with experimental data), are given elsewhere [62]. For example, the calculated crystal densities of some newly designed derivatives of tetrazole, namely, 5,5′-((1Z,5Z)-3,4-dinitrohexaaza-1,5-diene-1,6-diyl)bis(1-nitro-1H-tetrazole),5,5′-((1Z,5Z)-3,4-diaminohexaaza-1,5-diene-1,6-diyl)bis(1-nitro-1H-tetrazole),5,5′-((1Z,5Z)-3,4-dinitrohexaaza-1,5-diene-1,6-diyl)bis(1H-tetrazol-1-amine),3,3′-dinitro-3,3a,3′,3′a-tetrahydro-7H,7′H-[6,6′-bitetrazolo[1,5-e]pentazine]-7,7′-diamine,3,3′-,7,7′-tetranitro-3,3a,3′,3′a-tetrahydro-7H,7′H-6,6′-bitetrazolo[1,5-e]pentazine are 1.95, 1.84, 1.86, 1.90 and 1.92 g/cm³, respectively [35], which have high energy content [66]

Nirwan et al. [67] compared the reliability of the best available MESP-based methods to calculate the density of seven groups such as energetic compounds containing nitrate-esters, nitramines, and azides groups as well as energetic materials including benzene, caged, strained, heterocyclic backbone, and fused rings. They computed densities for 221 $C_aH_bN_cO_d$ explosives of different chemical nature and functional groups by these MESP-based methods and compared their outputs with the measured values. They indicated that Politzer et al. [22] as well as Rice and Byrd (Rice–Byrd) [62] methods can provide more reliable predictions as compared to the other MESP-based methods.

The crystal packing method accounts for the molecular interactions and conformation at a modest computational cost. It can also be used to predict the density of energetic compounds. A particular force field cannot be handled for all classes of energetic compounds. The option of force field depends on the type of atoms, hybridization, and chemical bonds in the desired molecule. Ghule and Nirwan [68] used various force fields in the crystal packing to evaluate their role for calculations of 68 energetic materials including aromatic and nonaromatic backbone with explosophoric groups. They predicted densities with the Dreiding force field for 84 % energetic compounds with the nonaromatic backbone that gave deviation within 5 %. For 83 % of energetic materials including nonaromatic backbone, the estimated densities with the polymer consistent force field (PCFF) were found within 4 % deviation. Thus, the selection of force field is essential for precise density estimation.

1.2.2 Energetic ionic liquids and salts as room temperature energetic materials

For ionic molecular crystals, Rice et al. [62] reparameterized the equation of Politzer et al. [22] as follows:

$$\rho = \alpha \frac{M}{V_m} + \beta \sum \left(\frac{\bar{V}_s^+}{A_s^+} \right) + \gamma \sum \left(\frac{\bar{V}_s^-}{A_s^-} \right) + \delta, \tag{1.8}$$

where the contributions from every ionic molecule in the formula unit are summed over the two ionic contribution terms. In this equation, $\bar{V}_s^+$ is the average of the positive values of V_s and A_s^+ is the portion of the cation surface that has a positive electrostatic potential. Furthermore, $\bar{V}_s^-$ is the average of the negative values of V_s and A_s^- is the portion of the anion surface that has a negative electrostatic potential. Rice et al. [62] obtained the values of α, β, γ and δ through best fit parameters and have the following values: 1.1145, 0.02056 g A$^{\circ 2}$/(cm^3 kcal/mol), −0.0392 g A$^{\circ 2}$/(cm^3 kcal/mol) and −0.1683 g/cm^3, respectively.

1.3 Empirical methods for the calculation of the crystal density of different classes of energetic materials

Empirical or quantitative structure-property relationship (QSPR) methods based on molecular structures are important because they are easy, effective, and precise methods for the density prediction of $C_aH_bN_cO_d$ explosives. The computer code EMDB_1.0 [69] can calculate densities of organic compounds containing common energetic functional groups including nitro (−NO$_2$), nitrate (−ONO$_2$), and nitramine (−NNO$_2$) using the best available QSPR methods. For some classes of energetic compounds, the reliability of these simple QSPR methods is higher than the best available group additivity approaches [5].

Some QSPR models based on complex molecular descriptors have also been developed in recent years. The molecular structures of 26 energetic cocrystals have been correlated with their densities by the artificial neural network (ANN) and multiple linear regression (MLR) analysis models [70] through three complex molecular descriptors. Two methods of ANN and MLR with five complex molecular descriptors have also been used to predict the density of 172 polynitroarenes, polynitroheteroarenes, nitroaliphatics, nitrate esters, and nitramines at room temperature [71]. It was found in these works that the ANN model can give a more reliable prediction as compared to the MLR model.

Different simple empirical methods have recently been introduced which enable the reliable prediction of the crystal density of important classes of energetic compounds at room temperature. These methods are reviewed here.

1.3.1 Nitroaromatic energetic compounds

It has been shown that the following general equation is suitable for most nitroaromatic high explosives [27]:

$$\rho' = \frac{10.57a + 0.1266b + 30.38c + 35.18d}{Mw},\tag{1.9}$$

where ρ' is the uncorrected crystal density in g/cm^3 (it can also be corrected for some compounds in which the molecular structure or intermolecular forces can result in a reduction or expansion of the volume of the compound), a, b, c and d are the number of carbon, hydrogen, nitrogen, and oxygen atoms respectively, and Mw is the molecular weight of the desired nitroaromatic energetic compound. This equation provides core correlation for estimation of the crystal density for a large number of nitroaromatic explosives. The corrected crystal density correlations can be expressed for some nitroaromatic energetic compounds according to the following:

(a) attachment of $-N_3$ or $-N_2^+$ groups to an aromatic ring,

$$\rho = -0.0238 + 0.9615\rho', \qquad (1.9a)$$

(b) presence of positive and negative charges on nitrogen:

$$\rho = 1.3022 + 0.3261\rho', \qquad (1.9b)$$

(c) existence of N-oxide in a heterocyclic aromatic structure:

$$\rho = 1.3958 + 0.2657\rho', \qquad (1.9c)$$

(d) attachment of an aromatic ring ($-Ar$) to $-OR$ or $-OAr'$ where R and Ar' are alkyl and aromatic groups, respectively:

$$\rho = -0.5139 + 1.2532\rho', \qquad (1.9d)$$

(e) the attachment of more than two $-OH$ or $-NH_2$ groups to the aromatic ring:

$$\rho = 1.1024\rho', \qquad (1.9e)$$

(f) attachment of one $-OH$ or two $-OH$, or two $-NH_2$ groups to an aromatic ring:

$$\rho = 0.2332 + 0.8872\rho', \qquad (1.9f)$$

(g) direct attachment of an aromatic ring to another aromatic ring (Ar–Ar):

$$\rho = -0.2164 + 1.093\rho'. \qquad (1.9g)$$

For any nitroaromatic compounds, the order in which the equations should be applied is from equations (1.9a)–(1.9g).

Example 1.8. Equation (1.9a) should be used rather than equation (1.9f) to calculate the crystal density of diazodinitrophenol (with the following molecular structure) because of the presence of the $-N_2^+$ group.

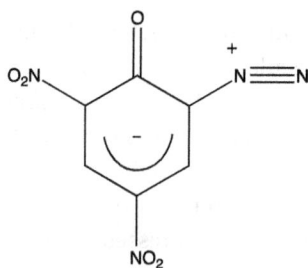

Thus, the crystal density of diazodinitrophenol ($C_6H_2N_4O_5$) is calculated as follows:

$$\rho' = \frac{10.57a + 0.1266b + 30.38c + 35.18d}{Mw}$$

$$= \frac{10.57(6) + 0.1266(2) + 30.38(4) + 35.18(5)}{210.1}$$

$$= 1.720 \text{ g/cm}^3$$

$$\rho = -0.0238 + 0.9615\rho'$$

$$= -0.0238 + 0.9615(1.720)$$

$$= 1.630 \text{ g/cm}^3.$$

The calculated value is the same as the measured value of 1.63 g/cm^3 [54].

1.3.2 Acyclic and cyclic nitramines, nitrate esters and nitroaliphatic compounds

A study of nitrate esters and nitroaliphatic systems shows that the atomic composition and number of special functional groups can be inserted into an empirical formula to predict the uncorrected crystal density of these compounds according to the following equation [26]:

$$\rho' = \frac{47.97a - 19.29b + 26.53c + 26.00d - 25.32n_{COO} - 0.6358n_O + 11.54n_{OH}}{Mw}, \quad (1.10)$$

where n_{COO}, n_O and n_{OH} are the number of ester, ether and alcohol functional groups, respectively.

The corrected crystal densities for some particular examples of acyclic and cyclic nitramines, nitrate esters, and nitroaliphatic compounds can be obtained based on the molecular structure as follows:

(1) mononitroalkanes:

$$\rho = 0.4170 + 0.5970\rho', \quad (1.10a)$$

(2) the attachment of two $-NO_2$ groups to one carbon (which has no additional $-COO-$, $-O-$ or $-OH$ functional groups):

$$\rho = 0.1233 + 0.8373\rho', \tag{1.10b}$$

(3) the attachment of three $-NO_2$ groups to one carbon:

$$\rho = \begin{cases} 3.033 - \rho' & \text{if } n_{CH_2} \geq 1.5 n_{NO_2}, \\ -0.3788 + 1.2569\rho' & \text{if } n_{CH_2} \leq 0.6 n_{NO_2}. \end{cases} \tag{1.10c}$$

For those molecules that satisfy both conditions (2) and (3), equation (1.10c) rather than (1.10b) should be used

(4) for nitrate compounds (without cyclic ring attachment), if $n_{CH_2ONO_2} + n_{CHONO_2} \geq 4$ then

$$\rho = 0.1745 + 0.9235\rho'; \tag{1.10d}$$

(5) cage and cyclo nitro compounds in which only one $-NO_2$ (not more) is attached to a carbon atom:

$$\rho = -0.0515 + 0.9142\rho'. \tag{1.10e}$$

For cyclic nitramines and nitramine compounds in which $N-NO_2$ is attached to an aromatic ring (such as in Tetryl), as well as polycyclic energetic compounds that contain no more than one oxygen atom in the ring, the crystal density can be calculated by

$$\rho = \frac{13.15a - 5.303b + 39.72c + 29.34d}{Mw}. \tag{1.11}$$

For acyclic nitramines, the following correlation is used:

$$\rho = \frac{66.86a - 27.37b + 52.96c + 12.81d}{Mw}. \tag{1.12}$$

Example 1.9. The calculated density of octanitrocubane ($C_8N_8O_{16}$) can be obtained as follows:

$$\rho' = \frac{47.97a - 19.29b + 26.53c + 26.00d - 25.32n_{COO} - 0.6358n_O + 11.54n_{OH}}{Mw}$$

$$= \frac{47.97(8) - 19.29(0) + 26.53(8) + 26.00(16) - 25.32(0) - 0.6358(0) + 11.54(0)}{464.1}$$

$$= 2.181 \text{ g/cm}^3$$

$$\rho = -0.0515 + 0.9142\rho'$$

$$= -0.0515 + 0.9142(2.181)$$

$$= 1.943 \text{ g/cm}^3.$$

The calculated value is close to the measured value which is reported to be 1.979 g/cm^3 [72].

1.3.3 Improved method for the prediction of the crystal densities of nitroaliphatics, nitrate esters and nitramines

For various nitroaliphatics, nitrate esters and nitramines, a more reliable general correlation for predicting the densities of acyclic and cyclic nitramines, nitrate esters and nitroaliphatic compounds than correlations (1.10), (1.11) and (1.12) was established, and can be written as follows [28]:

$$\rho = 1.521 + \frac{6.946a - 11.53b + 20.10c}{Mw} - 0.1559E_D + 0.1325E_I, \qquad (1.13)$$

where E_I and E_D are specific structural parameters that can increase or decrease the value of the crystal density, and the values of which are based on the molecular structure as follows:

(1) Nitroaliphatics and nitrate esters
  (a) $C(H)_{4-n}(NO_2 \text{ or } ONO_2)_n$: for nitro or nitrate derivatives of methane, the values of E_D and E_I are 1.70 and 0.0, respectively.
  (b) $C_nH_{2n+1}(NO_2 \text{ or } ONO_2)$: The values of E_D and E_I depend on the number of carbon atoms in the alkyl substituents of mononitro- or mononitrate-alkanes:
    (i) if $n = 2$, then $E_D = 1$ and $E_I = 0$;
    (ii) if $n = 3$, then $E_D = 0.5$ and $E_I = 0$;
    (iii) if $n = 4$, then $E_D = 0$ and $E_I = 0$;
    (iv) if $n = 5$, then $E_D = 0$ and $E_I = 0.5$;
    (v) if $n \geq 6$, then $E_D = 0$ and $E_I = 0.75$.
  (c) $C_nH_{2n}(NO_2 \text{ or } ONO_2)_2$: The values of E_D depend on the position of attachment of the nitro or nitrate groups in dinitro- or dinitrate-alkanes:
    (i) if two nitro or nitrate groups are attached to the same $-CH_2-$ group, $E_D = 1.5$ and $E_I = 0$;
    (ii) if two nitro or nitrate groups are attached to one $>CH-$ group, $E_D = 1.0$ and $E_I = 0$;
    (iii) for the other cases, $E_D = 0.75$ and $E_I = 0$.
  (d) If the OH group is present in nitro or nitrate compounds: $E_I = 0.4$ and $E_D = 0.0$.
  (e) Nitrate compounds without OH groups:
    (i) $E_I = 1.0$ and $E_D = 0$ for $(CH_2ONO_2)_4C$;
    (ii) $E_I = 0.5$ and $E_D = 0$ for compounds that contain $(CH_2ONO_2)_3$- or two $(CH_2ONO_2)_2$-fragments.
(2) Nitramines
  (a) $(C_nH_{2n+1})_2NNO_2$:
    (i) if $n = 1$, then $E_D = 1.5$ and $E_I = 0.0$;
    (ii) if $n = 2$, then $E_D = 0.5$ and $E_I = 0.0$;
    (iii) if $n = 3$, then $E_D = 0.0$ and $E_I = 0.0$;
    (iv) if $n \geq 4$, then $E_D = 0.0$ and $E_I = 0.75$.

(b) The presence of specific molecular entities:
    (i)   if an aromatic ring is present then $E_D = 0.5$ and $E_I = 0.0$;
    (ii) if the $NH-NO_2$ functional group is present then $E_D = 1.5$ and $E_I = 0.0$;
    (iii) for cyclic ethers $E_I = 1.0$ and $E_D = 0.0$;
    (iv) if the $-C(NO_2)_3$ group is present then $E_I = 0.5$ and $E_D = 0.0$.

Example 1.10. The value of the calculated density for $O_2NOCH_2CH_2ONO_2$ can be obtained as follows:

$$\rho = 1.521 + \frac{6.946a - 11.53b + 20.10c}{Mw} - 0.1559E_D + 0.1325E_I$$

$$= 1.521 + \frac{6.946(2) - 11.53(4) + 20.10(2)}{152.1} - 0.1559(0.75) + 0.1325(0)$$

$$= 1.457 \, \text{g/cm}^3.$$

It should be pointed out that this compound follows part (iii) of condition (1) (c). The estimated value is close to the measured value which is $1.48 \, \text{g/cm}^3$ [23].

1.3.4 Reliable correlation for the prediction of the crystal densities of polynitro arenes and polynitro heteroarenes

From studying the crystal densities of various polynitro arenes and polynitro heteroarenes, it has been shown that it is possible to correlate the crystal density with the elemental composition, as well as to establish the positive and negative contributions of some specific structural parameters using the following equation [29]:

$$\rho = -1.609 + \frac{29.20a + 1.515b + 53.06c + 61.30d}{Mw} + 0.0703C_{PG} - 0.0751C_{NG}, \quad (1.14)$$

where C_{PG} and C_{NG} are the positive and the negative contributions of some specific structural fragments, which can be specified according to the following.

Prediction of C_{PG}
(1) The presence of $n_{OH} \geq 1$ or $n_{NH_x} \geq 2$ without extra further substituents such as the methyl group:

$$C_{PG} = \begin{cases} 1.0 & \text{if } n_{NO_2} - n_{OH} = 1 \text{ or } n_{NO_2} - n_{NH_x} \geq 1; \\ 2.0 & \text{if } n_{NO_2} - n_{NH_x} = 0; \\ 0.5 & \text{if } n_{NO_2} - n_{OH} > 1. \end{cases}$$

(2) For polynitro benzene compounds containing a center of symmetry, or polynitro heteroarenes with substituent N-oxide or the explosive containing more than two

groups, $C_{PG} = 1.0$.

(3) For explosives that contain positive and negative charges on nitrogens, such as tetranitrodibenzo tetraazapentalene (TACOT), $C_{PG} = 2.0$.

(4) For polynitro arenes containing the ![structure] group, $C_{PG} = 3.0$.

Estimation of C_{NG}

(1) If the nitrate group is present:

$$C_{NG} = \begin{cases} 1.0 & \text{for } n_{NO_3} = 1 \\ 2.0 & \text{for } n_{NO_3} \geq 2. \end{cases}$$

(2) For polynitroaromatics with $-N_3$ or $-N_2$ substituents, or polynitro arenes containing more than two alkyl substituents, the value of C_{NG} is 1.0.
(3) If a polynitro arene cycle with only nitro substituents is directly attached to another $-Ar$, $-OR$ and $-OAr$ or $-NHAr$ (where R and Ar are alkyl and aromatic groups), the values of C_{NG} are 1.5, 1.0 and 0.75, respectively.
(4) For polynitro heteroarenes containing amino groups and in which more than two heteroatoms are present per ring, the value of C_{NG} is 1.0.

Example 1.11. 1,3,5-triazido-2,4,6-trinitrobenzene has the following structure:

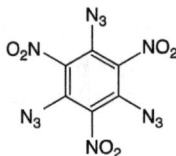

The use of equation (1.14) gives

$$\rho = -1.609 + \frac{29.20a + 1.515b + 53.06c + 61.30d}{Mw} + 0.0703C_{PG} - 0.0751C_{NG}$$

$$= -1.609 + \frac{29.20(6) + 1.515(0) + 53.06(12) + 61.30(6)}{336.1} + 0.0703(0) - 0.0751(1)$$

$$= 1.826 \text{ g/cm}^3.$$

The measured crystal density is $1.805\,\text{g/cm}^3$ [54]. Thus, the percent of deviation (%Dev) of the new method from the measured value is 1.24. The calculated crystal density which is obtained by the group additivity method of Ammon [24] is $1.630\,\text{g/cm}^3$ (%Dev = −9.70).

1.3.5 The extended correlation for the prediction of the crystal density of energetic compounds

It was found that it is possible to establish a general correlation to predict the crystal density of different classes of energetic compounds including various polynitroarenes, polynitroheteroarenes, nitroaliphatics, nitrate esters and nitramines as follows [73]:

$$\rho = 1.753 + \frac{-10.24b + 9.908c}{Mw} + 0.0992IMP - 0.0845DMP, \tag{1.15}$$

where IMP and DMP are two correcting functions that depend on intermolecular interactions for increasing and decreasing the second and third terms in equation (1.15), respectively. For different classes of energetic compounds, the values of IMP and DMP can be specified based on the molecular structure according to the rules which are outlined in the following subsections.

Structural parameters affecting *IMP*
(1) −OH or −ONH$_4$: The value of IMP is 1.0. This codition cannot be applied for the attachment of further alkyl groups to polynitroarenes.
(2) The attachment of both −NH$_x$ and nitro groups to carbocyclic aromatic molecular fragments: For those compounds that follow the conditions $1 \leq n_{NO_2} - n_{NH_x} < 0$ and $n_{NO_2} - n_{NH_x} = 0$, the values of IMP are 0.9 and 1.8, respectively.
(3) The presence of $\underset{+}{N}-\underset{-}{O}$, or more than two groups: If these molec-

ular fragments are present, the value of IMP is 1.0, except if the −NH− group is present between two aromatic rings, i. e. Ar−NH−Ar.
(4) Cyclic nitramines: The value of IMP is 1.0.
(5) (CH$_2$ONO$_2$)$_4$C and two groups (CH$_2$ONO$_2$)$_3$−: if the (CH$_2$ONO$_2$)$_4$C or two groups (CH$_2$ONO$_2$)$_3$− are present, the values of IMP are 1.5 and 0.5 respectively.
(6) Nonaromatic cyclic compounds containing nitro groups: The value of IMP equals 0.9.

Structural moieties affecting *DMP*
(1) Nitrobenzenes: For nitrobenzenes which contain only one benzene ring, $DMP = 0.5$, except for those compounds in which $b = 0$. If the two nitrobenzene rings are connected to each other through an −O− or −N=N− group, the values of DMP are 2.3 and 1.0, respectively.
(2) CH$_{4-n}$(NO$_2$ or ONO$_2$)$_n$ or the presence of the −N$_3$ group: The value of DMP is 3.3.
(3) C$_n$H$_{2n+1}$(NO$_2$ or ONO$_2$)$_2$:
   (a) if two nitro or nitrate groups are attached to the same carbon atom, $DMP = 2.2$;
   (b) if two nitro or nitrate groups are attached to two different carbon atoms, $DMP = 1.5$.
(4) Nonaromatic nitro and nitrate compounds in which −NO$_2$ or −ONO$_2$ groups are attached to a −CH$_2$− group: The value of DMP is 0.6 for the following conditions:
   (a) additional oxygen atoms are present in addition to those present in the −NO$_2$ or −ONO$_2$ groups in those energetic compounds with general formula R−CH$_2$−(NO$_2$ or ONO$_2$) or R−CH$_2$−(NO$_2$ or ONO$_2$)$_2$;
   (b) more than two −NO$_2$ or −ONO$_2$ groups are present.

Structural moieties affecting both *DMP* and *IMP*

(1) $C_{n+1}H_{2n+3}(NO_2$ or $ONO_2)$: The values of *IMP* and *DMP* depend on the number of carbon atoms in the alkyl substituents:
   (a) if $n = 1$, then $DMP = 1.9$;
   (b) if $n = 2$, then $DMP = 1.0$;
   (c) if $n = 3$, then $DMP = 0.0$;
   (d) if $n \geq 4$, then $IMP = 1.0$.
(2) $(C_nH_{2n+1})_2NNO_2$: The number of carbon atoms is important for the prediction of *IMP* and *DMP* in acyclic nitramines:
   (a) if $n = 1$, then $DMP = 2.1$;
   (b) if $n = 2$, then $DMP = 0.0$;
   (c) if $n \geq 3$, then $IMP = 1.4$.

This method is more complex than previous methods.

Example 1.12. 1,3,3-Trinitroazetidine (TNAZ) is a melt-cast explosive with the following molecular structure:

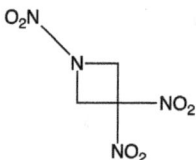

The present method can be applied as follows:

$$\rho = 1.753 + \frac{-10.24b + 9.908c}{Mw} + 0.0992IMP - 0.0845DMP$$

$$= 1.753 + \frac{-10.24(4) + 9.908(4)}{192.1} + 0.0992(1) - 0.0845(0)$$

$$= 1.845 \text{ g/cm}^3.$$

The predicted value is close to the measured value of 1.84 g/cm³ [72].

1.3.6 Energetic azido compounds

It has been shown that the molecular structures of energetic azido compounds can be used to evaluate the density using the following correlation [74]:

$$\rho = 1.2 + 0.01c + 0.26\frac{d}{a} - 0.1\frac{a}{c} + 0.04\rho_{azide}^+ - 0.52\rho_{azide}^-. \tag{1.16}$$

The two correcting parameters ρ^+_{azide} and ρ^-_{azide} are defined on the basis of the presence of certain molecular fragments as follows.

(1) The value of ρ^-_{azide} is 0.1 if two substituents of the following form are present:

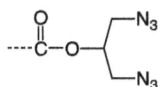

(2) The value of ρ^-_{azide} is 0.5 if the following molecular moiety is present in the compound, where R is an alkyl group:

(3) The value of ρ^+_{azide} is 1.0 if the following fragment is present:

(4) The value of ρ^+_{azide} is 2.0 if the $-C(NO_2)-CH_2-O-$ fragment is present.
(5) If more than two $-NNO_2$ groups are present, the values of ρ^+_{azide} are 0.5 and 2.0 for acyclic and cyclic nitramines, respectively.

Example 1.13. Using this method to calculate the density of 1,3-diazido-2-nitro-2-aza-propane with the following structure gives

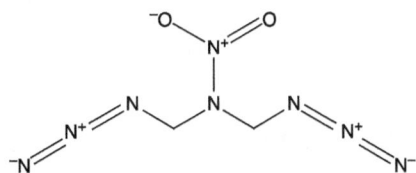

$$\rho = 1.2 + 0.01c + 0.26\frac{d}{a} - 0.1\frac{a}{c} + 0.04\rho^+_{azide} - 0.52\rho^-_{azide}$$

$$= 1.2 + 0.01(8) + 0.26\frac{2}{2} - 0.1\frac{2}{8} + 0.04(0) - 0.52(0)$$

$$= 1.52\,g/cm^3.$$

If the complex quantum mechanical method is used, a value of $1.45\,g/cm^3$ is obtained [74], while the experimental determined value is $1.43\,g/cm^3$ [75].

1.4 Empirical methods for the assessment of the crystal density of hazardous ionic molecular energetic materials using the molecular structures

Some QSPR methods based on the structure of ionic compounds have been introduced in recent years for predicting their densities. The computer code EMDB_1.0 [69] uses suitable QSPR methods to calculate densities of ionic molecular energetic materials. There are several empirical methods which can be used to predict the crystal density of ionic molecular energetic materials, and which are demonstrated here.

1.4.1 Two general empirical methods

It was shown that the elemental composition of an ionic molecular energetic compound with general formula $C_aH_bN_cO_d$ can be used to predict its crystal density as follows [76]:

$$\rho = 2.148 - \frac{7.767a + 6.261b + 4.154c}{Mw}. \tag{1.17}$$

The presence of some specific molecular moieties may enhance or reduce the molecular packing in ionic molecular energetic materials. Equation (1.17) can be improved by considering the effects of two correcting functions for increasing and decreasing density (ρ^+ and ρ^-) as

$$\rho = 2.137 - \frac{8.653a + 6.273b + 3.561c}{Mw} + 0.1241\rho^+ - 0.09772\rho^-, \tag{1.18}$$

where ρ^+ and ρ^- are two correcting functions that are equal to 1.0 for the presence of the following specific molecular moieties:

(1) ρ^+: the molecular fragments $>N\!-\!O^-$, $>N^-$, $\underset{}{\overset{}{\diagup}}\!N^+_{H_2}\diagup$ or $-\!N\!H^+\!<$ are present in a cyclic structure.

(2) ρ^-: the molecular moieties (N or)C$-$NH$^+$ and $>$N$-$CH$_3$ are present in a cyclic structure, as well as $\overset{(H\,or)C}{\underset{(H\,or)C}{>}}N-NH_3^+$, except in the presence of a bulky anion such as picrate.

Example 1.14. Triaminoguanidinium 4,5-dicyano-1,2,3-triazolate has the following structure, and the measured crystal density of this compound is 1.48 g/cm^3 [62]:

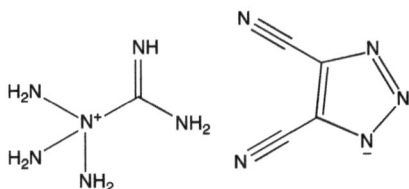

The use of equation (1.17) and (1.18) gives

$$\rho = 2.148 - \frac{7.767(5) + 6.261(9) + 4.154(11)}{223.2}$$

$$= 1.517 \text{ g/cm}^3 \quad (\%\text{Dev} = 2.5);$$

$$\rho = 2.137 - \frac{8.653(5) + 6.273(9) + 3.561(11)}{223.2} + 0.1241(0) - 0.09772(0)$$

$$= 1.515 \text{ g/cm}^3 \quad (\%\text{Dev} = 2.3).$$

The calculated values which have been obtained by two quantum mechanical methods reported by Rice et al. [20, 62] are 1.503 (%Dev = 1.5) and 1.428 g/cm³ (%Dev = −3.5), respectively.

1.4.2 The effects of various substituents on the density of tetrazolium nitrate salts

It was found that the effect of various substituents such as N_3, NO_2, NH_2, NF_2, CN and CH_3 on the density of tetrazolium nitrate salts with general formula $C_aH_bN_cO_dF_e$ can be shown by calculating the density using [77]:

$$\rho = 1.592 + 0.015c + 0.077e + 0.068P_{F_2} + 0.143P_{NO_2} - 0.081P_{CH_3}$$

$$- 0.072\rho_{\text{tetrazolium nitrate}}^-, \tag{1.19}$$

where P_{F_2}, P_{NO_2} and P_{CH_3} indicate the presence of the $-NF_2$, $-NO_2$, and $-CH_3$ groups in the salts, respectively. The value of $\rho_{\text{tetrazolium nitrate}}^-$ is 1.0 if the following structures are present:

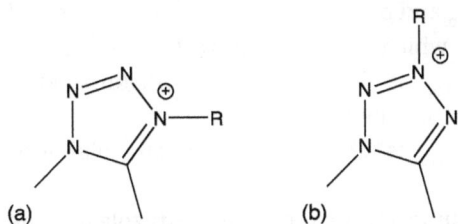

(a)   (b)

Example 1.15. The crystal density of the 1,5-dinitro-3H-tetrazolium nitrate salt containing the following cation

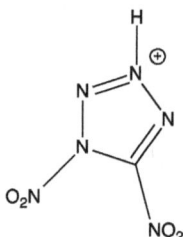

can be calculated as follows:

$$\rho = 1.592 + 0.015c + 0.077e + 0.068P_{F_2} + 0.143P_{NO_2} - 0.081P_{CH_3}$$
$$- 0.072\rho^-_{\text{tetrazolium nitrate}}$$
$$= 1.592 + 0.015(7) + 0.077(0) + 0.068(0) + 0.143(1) - 0.081(0)$$
$$- 0.072(0)$$
$$= 1.840 \text{ g/cm}^3.$$

The measured crystal density of this compound is 1.87 g/cm³ [37].

1.4.3 Predicting the density of tetrazole–N-oxide salts

A suitable correlation has been introduced to estimate the density of tetrazole N-oxide salts using molecular structure descriptors, which has the following form [78]:

$$\rho = 1.514 - 0.047a - 0.025b + 0.028c + 0.073d - 0.039n_{H_2O} + 0.075TET$$
$$+ 0.172\rho^+_{\text{Tetrazole–N-oxide}} - 0.118\rho^-_{\text{Tetrazole–N-oxide}} \qquad (1.20)$$

where n_{H_2O} is the number of the H_2O molecules in the crystal of the salt; the values of TET parameter are 1.0 and zero for tetrazole salts containing 1-N-oxide and 2-N-oxide fragment, respectively; $\rho^+_{\text{Tetrazole–N-oxide}}$ and $\rho^-_{\text{Tetrazole–N-oxide}}$ are the positive and negative nonadditive structural parameters, which are defined as [78]:

(a) $\rho^+_{\text{Tetrazole–N-oxide}}$: The value of $\rho^+_{\text{Tetrazole–N-oxide}}$ is 1.0 when there is a 2-hydroxytetrazole fragment in a tetrazole N-oxide energetic molecule.

(b) $\rho^-_{\text{Tetrazole–N-oxide}}$: The value of $\rho^-_{\text{Tetrazole–N-oxide}}$ is 1.0 for ammonium tetrazolate-1-N-oxide salts.

(c) For the presence of amino-nitroguanidinium as the cation of the tetrazole N-oxide salt, the value of $\rho^-_{\text{Tetrazole–N-oxide}}$ is 1.0.

Example 1.16. Bis(oxalyldihydrazidinium) 1H,1H-5,5-bitetrazole-1,1-diolate has the following molecular structure:

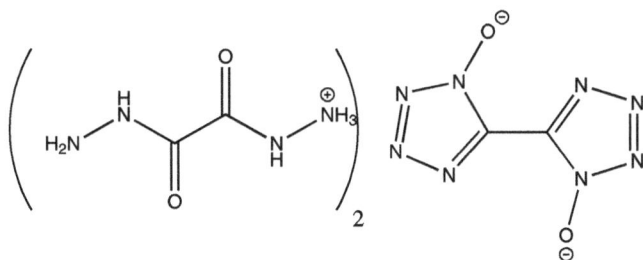

The use of equation (1.20) gives:

$$\rho = 1.514 - 0.047a - 0.025b + 0.028c + 0.073d - 0.039n_{H_2O} + 0.075TET$$
$$+ 0.172\rho^+_{\text{Tetrazole–N-oxide}} - 0.118\rho^-_{\text{Tetrazole–N-oxide}}$$
$$= 1.514 - 0.047 \times 6 - 0.025 \times 14 + 0.028 \times 16 + 0.073 \times 6 - 0.039 \times 0 + 0.075 \times 1$$
$$+ 0.172 \times 0 - 0.118\rho^-_{\text{Tetrazole–N-oxide}} \times 0$$
$$= 1.843 \, \text{g/cm}^3$$

The measured crystal density of this compound is 1.847 g/cm^3 [79].

1.5 Summary

Different empirical methods have been introduced in this chapter to predict the crystal densities of important classes of neutral and ionic molecular energetic compounds. Among different group additivity methods, more reliable approaches are introduced in section 1.1 for estimation of densities of neutral and ionic molecular energetic compounds at room temperature. For neutral energetic compounds, equations (1.9) to (1.12) are very simple approaches in comparison with other correlations, but their reliabilities are lower. Due to the high complexity of equation (1.15), it is recommended to use equations (1.13) and (1.14) to assess the crystal densities of various classes of energetic compounds, including polynitroarenes, polynitroheteroarenes, nitroaliphatics, nitrate esters, and nitramines with the general formula $C_aH_bN_cO_d$. For energetic azido compounds with general formula $C_aH_bN_cO_d$, equation (1.16) is a suitable relationship which can be used to predict the crystal density. For ionic molecular energetic compounds with general formula $C_aH_bN_cO_d$, the use of equation (1.18) is recommended to calculate the crystal densities. Finally, equations (1.18) and (1.19) provide good correlations for predicting the crystal densities of tetrazolium nitrate and tetrazole–N-oxide salts, respectively.

2 Heat of formation

The presence of certain energetic groups in organic compounds results in unstable compounds since the heats of detonation/combustion depend upon the presence of these functional groups [1, 10, 80]. Organic compounds containing energetic groups can decompose, ignite, or explode on exposure to external stimuli such as heat, impact, shock or an electric spark. Thus, the search for new organic energetic compounds with superior performance and lower sensitivity to undesired stimuli is important in both modern civil and military applications.

The condensed (solid or liquid) phase heat (or enthalpy) of formation of an energetic substance at 298.15 K is a measure of its energy content. It is an important factor to consider in the design, assessment and thermochemical stability of energetic compounds because knowledge of this value is essential in order to allow the evaluation of the performance properties of explosives and propellants such as the detonation pressure, heat of detonation, and specific impulse using theoretical methods [81–83]. Calorimetry is usually used to measure the condensed phase heat of formation of energetic compounds, but it is a time-consuming and destructive technique. Since it requires extremely pure samples [84], calorimetric measurements also have problems that can increase the uncertainty, e. g. completeness of the reaction or the production of undesired products [85]. Various computer codes which are used to calculate the detonation and combustion properties of energetic compounds such as EXPLO5 [86], EDPHT [87–89] and NASA-CEC-71 [90] require the value for the condensed phase heat of formation. Appendix A shows the measured values of the condensed phase heats of formation for well-known pure and composite explosives.

2.1 Condensed and gas phase heats of formation of energetic compounds

Different methods have been used to predict the condensed phase heat of formation of compounds containing energetic groups. Prediction of the condensed phase heat of formation can be achieved using different computational methods, in particular: group additivity, molecular mechanics (MM), quantum mechanical (QM) and empirical methods, or quantitative structure property relationships (QSPR) based on the

https://doi.org/10.1515/9783110740158-002

molecular structure [91, 92]. These methods can be classified in two different approaches for estimation of the condensed phase heat of formation. The first approach uses indirectly the heat of phase change property, i. e. the heats of sublimation and vaporization. The second approach utilizes directly the molecular structure of energetic materials wherein the methods of QSPRs were used. Many energetic compounds exist in the solid phase at room temperature. Thus, different methods were introduced to calculate the heat of sublimation in the first approach, i. e. group additivity [93, 94], QSPR based on complex descriptors [95], molecular surface electrostatic potential [96], and other QSPR schemes based on simple structural descriptors [97–100]. Many of the mentioned methods require complex computer codes but some models require the molecular structure of energetic materials, which are illustrated here.

2.1.1 QM, MM and group additivity approaches

Density functional theory (DFT) can provide highly accurate values for quantities such as bond strengths and heats of formation of energetic compounds. Sana et al. [101] introduced the term stabilization energy, which measures special effects such as bond interaction and electron dislocation. It was shown that $-NO_2$ groups result in a high destabilization energy. Different QM methods can predict the gas phase heats of formation of energetic compounds [81]. The determination of the geometries and enthalpies of formation of molecules using MM is widely used for large chemical systems. Thus, MM2 [102], MM3 [103] and MM4 [104] have been parametrized [105]. The MM2 method has been used for the prediction of the gas phase heats of formation of nitro compounds [106]. MM methods are fast and inexpensive, however, due to the lack of reliable parameters for some compounds, their application is limited [84, 107]. Several semiempirical QM methods have been parametrized to enable the prediction of the energies and enthalpies of formation. In contrast to MM methods, parametrization of the semiempirical methods has been performed over a wide range of atoms and compounds. A more general parametrization leads to an increase in the uncertainties of the semiempirical methods [84]. Thus, several modifications such as AM1 [108], PM3 [109], PM6 [110] and PM7 [111] have been introduced to extend the domain, as well as to increase the accuracy. AM1 and PM3 methods have been used to predict the gas phase heats of formation of aromatic and aliphatic nitro compounds [112]. The PM3 method has also been used for the indirect prediction of the condensed phase heats of formation of nitroaromatic compounds [113].

Hess's law can be used to calculate the solid and liquid phase heats of formation using the predicted heats of vaporization and sublimation. The heat of sublimation or sublimation enthalpy has an important role in the assessment of the strength of intermolecular cohesion. It provides a valuable indicator of crystal stability and vapor pressure of organic compounds containing energetic groups such as $-O-O-$, $-N_3$, $-ON=O$, $-NO_2$, $-ONO_2$, and $-NNO_2$ in the solid state [114]. Energetic compounds may exist in the solid or liquid phase at room temperature. Solid and liquid phase heats of formation of energetic compounds can be obtained by subtraction of the heat of sublimation from the gas phase value as [115]:

$$\Delta_f H^\theta(s) = \Delta_f H^\theta(g) - \Delta_{sub} H^\theta \qquad (2.1)$$

$$\Delta_f H^\theta(l) = \Delta_f H^\theta(g) - \Delta_{vap} H^\theta, \qquad (2.2)$$

where $\Delta_f H^\theta(s)$ is the standard solid phase heat of formation; $\Delta_f H^\theta(l)$ is the standard liquid phase heat of formation; $\Delta_f H^\theta(g)$ is the standard gas phase heat of formation; $\Delta_{sub} H^\theta$ is the standard heat of sublimation; $\Delta_{vap} H^\theta$ is the standard heat of vaporization. Some accurate methods exist for the calculation of $\Delta_f H^\theta(g)$ for neutral energetic compounds. Byrd and Rice [116] used DFT coupled with an atom and group contribution method for estimation of $\Delta_f H^\theta(g)$. For a test set of 45 energetic compounds, Ohlinger et al. [117] introduced T1 as a novel multilevel computational method to compare the ability of six different methods for accurate calculation of $\Delta_f H^\theta(g)$. Akutsu et al. [106] combined the heats of vaporization and sublimation using the additivity rule with the calculated gas phase heats of formation data from the PM3 and MM2 methods to calculate the condensed phase heats of formation. It was shown that there is a correlation between the statistically-based quantities of electrostatic potentials mapped onto isodensity surfaces of isolated molecules and their heats of sublimation and vaporization [118]. Rice and coworkers [119, 120] used the 6-31G* basis set and the hybrid B3LYP density functional to convert quantum mechanical energies of molecules into gas phase heats of formation. They used the surface electrostatic potentials of individual molecules for computation of the heats of sublimation and vaporization.

Politzer et al. [121] introduced the following equation for calculation of the enthalpy of sublimation:

$$\Delta_{sub} H^\theta = w_1(SA)^2 + w_2(\sigma_{tot}^2 \nu)^{0.5} + w_3, \qquad (2.3)$$

where SA is molecular surface area; σ_{tot}^2 indicates the variability of the potential on the molecular surface; ν shows a degree of the balance between positive and negative regions. The mentioned parameters w_1, w_2, and w_3 in equation (2.3) can be determined from least-squares fitting to reliable values of the enthalpies of sublimation of organic compounds containing energetic groups. Byrd and Rice [116] used experimental data $\Delta_{sub} H^\theta$ for 23 different energetic compounds to obtain these parameters. Equation (2.3) was widely used in recent years for the prediction of the $\Delta_f H^\theta(s)$ values of newly designed energetic compounds [122–128]. Some efforts have been made to improve the prediction accuracy of equation (2.3) by optimizing the parameters w_1, w_2, and w_3 for a very narrow class of compounds such as energetic tetrazine derivatives [129].

Politzer et al. [121] inserted an average deviation of electrostatic potential (Π) in equation (2.3) to measure local polarity as an additional parameter:

$$\Delta_{sub} H^\theta = w_1(SA)^2 + w_2(\sigma_{tot}^2 \nu)^{0.5} + w_3 + w_4 \Pi. \qquad (2.4)$$

Suntsova and Orofeeva [96] adjusted the parameters w_1, w_2, w_3 and w_4 in equations (2.3) and (2.4) using the reported enthalpies of sublimation for 185 compounds, including 148 and 37 compounds for training and test sets, respectively. The overall performances of equations (2.3) and (2.4) are close to each other but the maximum absolute deviation of equation (2.4) is lower [96]. Equation (2.4) was used to predict $\Delta_f H^\theta(s)$ some tetrazole-, tetrazine-, furazan-, and furoxan-based energetic compounds [96].

Group additivity methods can be used to estimate the ideal gas phase heats of formation, using, for example, the methods of Benson, Yoneda and Joback [130]. For these methods, assemblies of adjacent atoms are defined as groups, so that the enthalpy of the molecule is calculated by the summation of the contributions of the groups. Further parameters may be used to consider the effects of strain, resonance and conjugation. Group additivity methods can also be used to calculate the condensed phase heat of formation for some specific classes of energetic compounds. Due to the presence of different molecular interactions, molecular packing, and polymorphism in some solid compounds, prediction of the condensed phase heat of formation is more difficult than of the gas phase heat of formation [131]. Several group additivity methods have been used to predict the condensed phase heat of formation of common CHNO energetic materials [131, 132]. For example, Bourasseau [133] applied the group additivity method to predict the standard heats of formation at 298 K of aliphatic and alicyclic polynitro compounds. Salmon and Dalmazzone [132] also introduced a group contribution method that can be applied to large classes of CHNO energetic compounds to predict enthalpies of formation in the solid state (at 298.15 K). Argoub et al. [131] introduced a suitable method which provides significant improvements in accuracy and applicability as compared to the various group contribution methods for estimating $\Delta_f H^\theta(s)$. Their model has a simple linear form as [131]:

$$\Delta_f H^\theta(s) = \sum_i^N n_i \times A_i, \tag{2.5}$$

where A_i shows the contribution of the first-, second-, or third-order group of type i, which occurs n_i times [134] in a desired organic compound. Reliability of group contribution methods is lower than that of quantum mechanical methods as well as of QSPR models based on molecular moieties for estimation of $\Delta_f H^\theta(s)$ in high-energy content organic compounds [88, 134–138]. It was shown that the reliability of the method of Argoub et al. [131] is lower than that of QSPR models based on molecular fragments [134]. However, group additivity methods have some restrictions, e. g.
(1) they cannot be used for energetic materials which exist in the liquid state at 298.15 K,
(2) the group contributions of some functional groups have not yet been defined in the models, and
(3) for compounds which exhibit unusual chemical structures, additivity methods cannot be used.

2.1.2 Empirical approaches or QSPR methods on the basis of structural parameters

There are several methods which can be used to predict the condensed phase heats of formation of some classes of energetic compounds. Among these methods, there is a complex method in which the solid phase heat of formation of a desired CHNO explosive in the range $Q_{corr} > 4602\,kJ/g$ [139] (Q_{corr} is the corrected heat of detonation on the basis of Kamlet's method [140]), can be predicted based on its approximate detonation temperature. The other methods use the molecular structures of energetic compounds, and are reviewed here.

Simple procedure for nitroaromatic energetic materials
It was shown that the following general equation is suitable for calculating the condensed phase heat of formation for most nitroaromatic and benzofuroxan-based energetic compounds with general formula $C_aH_bN_cO_d$ [141]:

$$\Delta_f H^\theta(c) = 32.76a - 33.96b + 69.12c - 116.32d$$
$$+ 124.8n_{NO_2} - 65.10n_{Ar-NH} - 93.64n_{OH} - 202.3n_{COOH}$$
$$+ 13.56(n_{Ar} - 1) + 121.4n_{-N=N-} + 223.2n_{cyclo-N-O-N}, \tag{2.6}$$

where $\Delta_f H^\theta(c)$ is the standard heat of formation of a specific compound in the condensed phase (solid or liquid) in kJ/mol; n_{NO_2}, n_{OH} and n_{COOH} are the number of nitro, hydroxyl, and carboxyl functional groups, respectively; n_{Ar-NH} is the number of NH (or NH_2) functional groups attached to aromatic rings, n_{Ar} is the number of aromatic rings, $n_{-N=N-}$ is the number of non-cyclic $-N=N-$ groups and $n_{cyclo-N-O-N}$ is the number

of $\left[\begin{smallmatrix}-N\\ \\ -N\end{smallmatrix}\!\!>O\right.$ (in benzofuroxan compounds) groups.

Example 2.1. The condensed phase heat of formation of

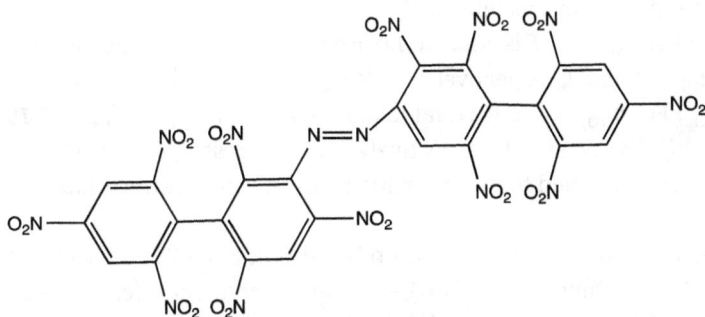

is calculated as follows:

$$\Delta_f H^\theta(c) = 32.76a - 33.96b + 69.12c - 116.32d$$
$$+ 124.8n_{NO_2} - 65.10n_{Ar-NH} - 93.64n_{OH} - 202.3n_{COOH}$$

$$+ 13.56(n_{Ar} - 1) + 121.4n_{-N=N-} + 223.2n_{cyclo-N-O-N}$$
$$= 32.76(24) - 33.96(6) + 69.12(14) - 116.32(24)$$
$$+ 124.8(12) - 65.10(0) - 93.64(0) - 202.3(0)$$
$$+ 13.56(4 - 1) + 121.4(1) + 223.2(0)$$
$$= 417.8 \text{ kJ/mol.}$$

The measured $\Delta_f H^\theta(c)$ of this compound is 480.3 kJ/mol [142].

More reliable approach for nitroaromatic energetic materials

It was shown that a more reliable approach can be used to calculate the condensed phase heat of formation for nitroaromatic compounds that have complex and different molecular structures according to [137]:

$$\Delta_f H^\theta(c) = \frac{\begin{array}{c} 2.690a - 2.896b + 2.876c - 2.784d \\ -1.701(n'_{Ar} - 1) - 1.607(\frac{n_{NO_2}}{n_{DFG/SP}} \times E) + 3.246(\frac{n_{IFG/SP}}{n_{NO_2}} \times F) \end{array}}{Mw \times 10^{-4}}, \qquad (2.7)$$

where Mw is the molecular weight of nitroaromatic compound. Three variables n'_{Ar}, E and F can be predicted based on the following situations:

(1) n'_{Ar}: The value of n'_{Ar} is equal to n_{Ar}. Therefore, for the presence of one and two aromatic rings, n'_{Ar} is equal to one and two, respectively. In the case of $n_{Ar} \geq 3$, n'_{Ar} equals n_{Ar} except if there is one nitro group (e. g. in 2,2',2'',4,4',4'',6,6',6''-nona-nitro-1,1':3',1''-terphenyl) or nitrogen atom (e. g. in 2,4,6-tripicryl-1,3,5-triazine) between two aromatic rings in which $n'_{Ar} = 0$. If the $-N=N-$ group is attached to an aromatic ring, the value of n'_{Ar} is also equal to zero.

(2) $n_{DFG/SP}$ and E: The ratio $n_{NO_2}/n_{DFG/SP}$ and E are defined for different functional groups or structural parameters as follows:

(a) $n_{DFG/SP} = n_{OH}$: For $n_{NO_2} = 1$, E is equal to 1.0. However, if $n_{NO_2} > 1$, then $E = 0.5$ and 1.75 for $n_{OH} = 1$ and 2, respectively.

(b) $n_{DFG/SP} = n_{NH_x}$: For $n_{NO_2} = 1$, E is equal to 0.0. However, if $n_{NO_2} > 1$, $E = 0.75$ and 0.667 for $n_{NH_x} = 1$ and > 1, respectively. For the presence of a ring containing $>$NH group attached to two aromatic rings (e. g. tetranitrocarbazole), $E = 1.25$.

(c) $n_{DFG/SP} = n_{-C(=O)-}$: If the $-C(=O)-OH$ group is present, $E = 1.75$. The value of $E = 0.875$ for the attachment one $-C(=O)-$ to two aromatic rings (e. g. (2,4-di-nitrophenyl)(2,4,6-trinitrophenyl)methanone).

(d) Polynitro naphthalene: $n_{NO_2}/n_{DFG/SP} \times E = 2.0$.
The higher value of $n_{NO_2}/n_{DFG/SP} \times E$ can be used for the presence of multiple types of functional groups, e. g. $-OH$ and $-NH_2$. Thus, $n_{NO_2}/n_{DFG/SP} \times E = 0.75$ in 2-amino-4,6-dinitro-phenol.

(3) $n_{IFG/SP}$ and F: The ratio $n_{IFG/SP}/n_{NO_2}$ can be used to predict the values of F as follows:

(a) $n_{IFG/SP} = n_{-R/-OR}$: on attachment of $-R$ or $-OR$ to an aromatic ring, $F = 2.0$.

(b) $n_{IFG/SP} = n_{-NH-NH_2}$: on attachment of hydrazine to an aromatic ring, $F = 1.0$.

(c) $n_{IFG/SP} = n_{-N=N-}$: If the ratio $n_{-N=N-}/n_{NO_2} \geq 0.167$, $n_{NO_2}/n_{DFG/SP} \times F = 12.0$. For other nitro compounds containing $-N=N-$ groups, $n_{NO_2}/n_{DFG/SP} \times F = 6.0$.

Example 2.2. The condensed phase heat of formation of the compound given in Example 2.1 is calculated as

$$\Delta_f H^\theta(c) = \frac{\begin{array}{l} 2.690a - 2.896b + 2.876c - 2.784d \\ -1.701(n'_{Ar} - 1) - 1.607(\frac{n_{NO_2}}{n_{DFG/SP}} \times E) + 3.246(\frac{n_{IFG/SP}}{n_{NO_2}} \times F) \end{array}}{Mw \times 10^{-4}}$$

$$= \frac{\begin{array}{l} 2.690(24) - 2.896(6) + 2.876(14) - 2.784(24) \\ -1.701(0) - 1.607(0) + 3.246(6) \end{array}}{874.4 \times 10^{-4}}$$

$$= 458.7 \text{ kJ/mol}.$$

Since the measured $\Delta_f H^\theta(c)$ of this compound is 480.3 kJ/mol [142], the predicted $\Delta_f H^\theta(c)$ result obtained using equation (2.7) is close to the value obtained from the experimental data.

Using the estimated gas phase enthalpies of formation from the PM3 and B3LYP methods

For nitroaromatic energetic compounds, it was found that the estimated gas phase heat of formation can be used to predict the condensed phase heat of formation. Two suitable correlations on the basis of the B3LYP/6-31G* and PM3 as follows [113]:

$$\Delta_f H^\theta(c) = 0.874[\Delta_f H^\theta(g)]_{B3LYP/6\text{-}31G*} + 35.575a - 22.59b - 31.947d$$
$$+ 30.5n_{NO_2} - 141.91n_{Ar}, \tag{2.8}$$

$$\Delta_f H^\theta(c) = 0.911[\Delta_f H^\theta(g)]_{PM3} - 10.8b + 26.968d$$
$$+ 46.17n_{NO_2} - 101.14n_{N_2} - 319.129n_{TRs}, \tag{2.9}$$

where $[\Delta_f H^\theta(g)]_{B3LYP/6\text{-}31G*}$ and $[\Delta_f H^\theta(g)]_{PM3}$ are the calculated gas phase heats of formation in kJ/mol using the B3LYP/6-31G* and PM3 methods, respectively; n_{N_2} is the number of $-N=N-$ or $-N\equiv N$ groups; the n_{TRs} is the number of the attachment of three aromatic rings. These correlations are more complex than equations (2.6) and (2.7) because they require quantum mechanical computations.

Simple method for nitramines, nitrate esters and nitroaliphatics

From studying various nitramines, nitrate esters, nitroaliphatics and related energetic compounds, it could be shown that the following equation can provide a suitable pathway for obtaining $\Delta_f H^\theta(c)$ [143]:

$$\Delta_f H^\theta(c) = 29.68a - 31.85b + 144.2c - 90.71d$$

$$- 88.84 n_{OH} - 39.14 n_{N-NO_2} - 45.62 n_{>C=O} + 256.3 n_1^0$$

$$- 380.5 n_{=C<_N^N} + 30.20 n_{O-NO_2}, \tag{2.10}$$

where n_{OH}, n_{N-NO_2}, $n_{>C=O}$ and n_{O-NO_2} are the number of specified functional groups, $n_1^0 = 0$ if hydrogen is present in the molecule, and $n_1^0 = 1$ for hydrogen-free compounds; $n_{=C<_N^N}$ is the number of $=C<_N^N$ structural moieties present in the energetic compound.

Example 2.3. The condensed phase heat of formation of the following compound

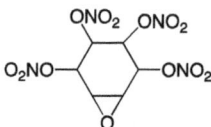

is calculated as

$$\Delta_f H^\theta(c) = 29.68a - 31.85b + 144.2c - 90.71d$$

$$- 88.84 n_{OH} - 39.14 n_{N-NO_2} - 45.62 n_{>C=O} + 256.3 n_1^0$$

$$- 380.5 n_{=C<_N^N} + 30.20 n_{O-NO_2}$$

$$= 29.68(6) - 31.85(6) + 144.2(4) - 90.71(13)$$

$$- 88.84(0) - 39.14(0) - 45.62(0) + 256.3(0)$$

$$- 380.5(0) + 30.20(4)$$

$$= -494.6 \text{ kJ/mol.}$$

The measured $\Delta_f H^\theta(c)$ of this compound is -444.3 kJ/mol [142].

More reliable method for acyclic and cyclic nitramines, nitrate esters and nitroaliphatic energetic compounds

It was shown that a more reliable correlation than that of equation (2.10) can be established, which is based on the molecular structure of energetic compounds. This correlation can be formulated as follows [136]:

$$\Delta_f H^\theta(c) = 39.10a - 37.89b + 96.73c - 66.07d$$

$$+ 215.4 \sum_i n_{i,DE} - 217.6 \sum_j n_{j,IE}. \tag{2.11}$$

The factors n_{DE} and n_{IE} are the number of some specific functional groups or structural parameters, which may decrease or increase the value of $\Delta_f H^\theta(c)$, respectively. The values of n_{DE} and n_{IE} can be determined as follows.

(1) Prediction of n_{DE}

    (a) −OH group: If the ratio of the number of hydroxyl groups to the number of nonaromatic nitro or nitrate groups $(n_{OH}/n_{NO_2\ or\ ONO_2}) \geq 1$, then $n_{DE} = n_{OH}/n_{NO_2\ or\ ONO_2} \times 0.5$ (e. g. for 2-hydroxymethyl-2-nitro-propane-1,3-diol, $n_{OH}/n_{NO_2\ or\ ONO_2} = 3$ and $n_{DE} = 1.5$). For $n_{OH}/n_{NO_2\ or\ ONO_2} < 1$, $n_{DE} = 0.25$ (e. g. for 2,2,2-trinitro-ethanol, $n_{OH}/n_{NO_2\ or\ ONO_2} = 0.33$).

    (b) >C=O group: For compounds containing the >C=O group, $n_{DE} = n_{>C=O} \times 0.6$ (e. g. for 4,4,4-trinitro-butyric acid 2,2,2-trinitro-ethyl ester, $n_{DE} = 0.6$).

    (c) Cyclic and acyclic ether functional groups: For six-membered cyclic ether rings only, $n_{DE} = 0.25$. If there is an ether functional group of the type $ROCH_2CH_2OR'$, $n_{DE} = 0.5$ (e. g. 1-nitrooxy-2-[2-(2-nitrooxy-ethoxy)-ethoxy]-ethane). The value of n_{DE} is found to be 0.25 for other types of acyclic ethers (e. g. 1-nitrooxy-2-(2-nitrooxy-ethoxy)-ethane).

    (d) Some specific molecular structures: if the ![fragment] , ![fragment with NO2] or −NH−NO₂ molecular fragments are present, the values of n_{DE} correspond to 1.0, 0.5 and 0.2, respectively. For example, there are two ![fragment] and one −NH−NO₂ molecular fragments in 1-nitro-3-guanidinourea which gives $\sum_i n_{DE} = 2 + 0.2 = 2.2$.

(2) Prediction of n_{IE}

    (a) Acyclic and cyclic nitramines containing only one C–N(NO₂)–C fragment: The value of $n_{IE} = 0.3$ (e. g. N-ethyl-N-nitro-ethanamine).

    (b) The number of nitro groups attached to cubane (C_8H_8): The value of n_{IE} is 0.2. For example, $\sum_i n_{IE} = 8 \times 0.2 = 1.6$ for octanitrocubane.

    (c) Hydrogen-free nitroalkanes: The value of $n_{IE} = 1.0$ (e. g. tetranitometane).

Example 2.4. The condensed phase heat of formation of the compound given in Example 2.3 is calculated as

$$\Delta_f H^\theta(c) = 39.10a - 37.89b + 96.73c - 66.07d$$
$$+ 215.4 \sum_i n_{i,DE} - 217.6 \sum_j n_{j,IE}$$
$$= 39.10(6) - 37.89(6) + 96.73(4) - 66.07(13)$$
$$+ 215.4(0) - 217.6(0)$$
$$= -464.7\ \text{kJ/mol.}$$

Since the measured $\Delta_f H^\theta(c)$ of this compound is −444.3 kJ/mol [142], the predicted $\Delta_f H^\theta(c)$ obtained using equation (2.9) is close to the value obtained from experimental data.

Prediction of the condensed phase heats of formation of polynitro arenes, polynitro heteroarenes, acyclic and cyclic nitramines, nitrate esters and nitroaliphatic compounds

For polynitro arenes, polynitro heteroarenes, acyclic and cyclic nitramines, nitrate esters, and nitroaliphatic compounds, the following correlation can be used to predict the condensed phase heat of formation [138]:

$$\Delta_f H^\theta(c) = 32.33a - 39.49b + 92.41c - 63.85d$$
$$+ 105.0\Delta_f H^\theta_{IEC} - 106.6\Delta_f H^\theta_{DEC}, \tag{2.12}$$

where $\Delta_f H^\theta_{IEC}$ and $\Delta_f H^\theta_{DEC}$ are two correcting functions which can be specified based on the presence or absence of some groups, which are described in the following situations.

(1) Prediction of $\Delta_f H^\theta_{DEC}$:

   (a) −OH group: The values of $\Delta_f H^\theta_{DEC}$ are 1.4 and 1.0 for energetic compounds containing the −OH group as Ar−OH and R−OH, respectively (e. g. $\Delta_f H^\theta_{DEC} = 2 \times 1.4 = 2.8$ for 2,4,6-trinitrobenzene-1,3-diol).

   (b) −NH$_x$ groups: If the −NH$_2$ or >NH (or −NH−NH$_2$) group is present in the energetic compound, the value of $\Delta_f H^\theta_{DEC}$ is equal to 0.7 (e. g. $\Delta_f H^\theta_{DEC} = 3 \times 0.7 = 2.1$ for 2,4,6-trinitrobenzene-1,3,5-triamine).

   (c) Acyclic and cyclic ether functional groups: The values of $\Delta_f H^\theta_{DEC}$ are 0.5 and 0.9 (except cyclic ethers with three-membered rings), respectively (e. g. $\Delta_f H^\theta_{DEC} = 1 \times 0.5 = 0.5$ for 2-methoxy-1,3,5-trinitrobenzene).

   (d) Other specific polar groups: If the −COOH (or −O$^-$NH$_4^+$, −COCO− and −NH−CO−), −N−CO−N−, −COO− (or acyclic −CO−), −CO−H, cyclic-CO− or −NH−NO$_2$ functional groups are present, the contributions of these groups to $\Delta_f H^\theta_{DEC}$ correspond to 2.8, 2.4, 1.4, 1.0, 0.5 and 0.3, respectively (e. g. $\Delta_f H^\theta_{DEC} = 1 \times 2.8 = 2.8$ for 3,5-dinitrobenzoic acid).

   (e) The number of nitrogen heteroatoms in six-membered rings: For nitroaromatics containing more than one six-membered aromatic ring, the contribution of each nitrogen heteroatom in the six-membered ring is 0.33 (e. g. $\Delta_f H^\theta_{DEC} = 3 \times 0.33 = 0.99$ for 2,4,6-tris(2,4,6-trinitrophenyl)-1,3,5-triazine)

(2) Prediction of $\Delta_f H^\theta_{IEC}$:

   (a) Cyclic and acyclic nitramines: The values of $\Delta_f H^\theta_{IEC}$ are 0.5 and 0.8 for the acyclic and cyclic functional groups, respectively (e. g. $\Delta_f H^\theta_{IEC} = 0.5$ for 1,3,5-trinitro-1,3,5-triazinane).

   (b) R−NO$_2$ (or R−ONO$_2$) and Ar−N$_3$: The value of $\Delta_f H^\theta_{IEC}$ is 0.3 for the attachment of an −NO$_2$ or −ONO$_2$ group to a nonaromatic carbon atom, and for the attachment of the −N$_3$ group to an aromatic ring, (e. g. $\Delta_f H^\theta_{IEC} = 0.3$ for 1,1-dinitropropane). There are some exceptions:

      (i) if the −C(NO$_2$)$_3$ group is present, $\Delta_f H^\theta_{IEC} = 0.8$ (e. g. $\Delta_f H^\theta_{DEC} = 1.0$ and $\Delta_f H^\theta_{IEC} = 0.8$ for 2,2,2-trinitroethanol);

(ii) for linear mono-nitroalkanes ($a \geq 4$), $\Delta_f H_{IEC}^{\theta} = 0.6$ (e. g. $\Delta_f H_{IEC}^{\theta} = 0.6$ for 1-nitrobutane);

(iii) for hydrogen-free nitroalkanes, $\Delta_f H_{DEC}^{\theta} = 2.2$ (e. g. $\Delta_f H_{IEC}^{\theta} = 2.2$ for tetra-nitromethane);

(iv) for symmetric linear di-nitroalkanes, $\Delta_f H_{IEC}^{\theta} = 0.0$ (e. g. $\Delta_f H_{DEC}^{\theta} = 0.0$ for 1,2-dinitroethane).

(c) $-N=N-$ and $\overset{\diagdown}{\underset{+}{N}}-\overset{-}{O}$: For the molecular fragments $\overset{\diagdown}{\underset{+}{N}}-\overset{-}{O}$ and $-N=N-$, the contributions to $\Delta_f H_{IEC}^{\theta}$ are 0.7 and 0.8, respectively ($\Delta_f H_{IEC}^{\theta} = 3 \times 0.7 = 2.1$ for benzenetrifuroxan).

(d) Number of carbocyclic aromatic rings ($n_{Ar,car}$): For energetic compounds which only contain carbocyclic aromatic rings, the contribution to $\Delta_f H_{IEC}^{\theta}$ is ($n_{Ar,car} - 1$) $\times 0.3$ (e. g. $\Delta_f H_{IEC}^{\theta}$ is (2 – 1) $\times 0.3 = 0.3$ for 2,2′,4,4′,6,6′-hexanitrobiphenyl).

(e) Attachment of alkyl groups to an aromatic ring: The values of $\Delta_f H_{IEC}^{\theta}$ are 0.2 and 0.8 for the attachment of methyl and longer carbon chain alkyl groups (or $-CH=CH-$) to aromatic rings, respectively (e. g. $\Delta_f H_{IEC}^{\theta} = 0.2$ for 1-methyl-2,4-dinitrobenzene). If $n_{Ar,car} > 1$, it is not necessary to include condition (d).

(f) Nitro groups attached to non-aromatic four-membered rings: $\Delta_f H_{IEC}^{\theta} = 0.45 \times$ the number of $-NO_2$ groups attached to a non-aromatic four-membered ring (e. g. $\Delta_f H_{IEC}^{\theta} = 8 \times 0.45$ for octanitrocubane).

Example 2.5. The value of $\Delta_f H^{\theta}(c)$ for the following molecular structure is calculated as

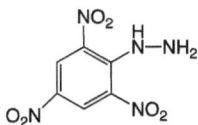

$$\Delta_f H^{\theta}(c) = 32.33a - 39.49b + 92.41c - 63.85d + 105.0\Delta_f H_{IEC}^{\theta} - 106.6\Delta_f H_{DEC}^{\theta}$$

$$= 32.33(6) - 39.49(5) + 92.41(5) - 63.85(6) + 105.0(0) - 106.6(0.7)$$

$$= 0.8 \text{ kJ/mol.}$$

The measured $\Delta_f H^{\theta}(c)$ of this compound is 36.5 kJ/mol [142].

General correlation for organic energetic materials

It has been established that the following general correlation can be used for organic energetic compounds containing different types of energetic groups such as $-NO_2$,

$-ONO_2$, $-NNO_2$, $-ON=O$, $-O-O-$ and $-N_3$ [144]:

$$\Delta_f H^\theta(c) = -111.4 + 33.11a - 28.84b + 86.80c - 62.03d$$
$$- 79.25(\Delta_f H^\theta_{add,DHC} + \Delta_f H^\theta_{nonadd,DHC})$$
$$+ 153.3(\Delta_f H^\theta_{add,IHC} + \Delta_f H^\theta_{nonadd,IHC}), \tag{2.13}$$

whereby the subscripts DHC and IHC in the $\Delta_f H^\theta_{add,DHC}$, $\Delta_f H^\theta_{nonadd, DHC}$, $\Delta_f H^\theta_{add,IHC}$ and $\Delta_f H^\theta_{nonadd, IHC}$ show decreasing and increasing heat contents in the compounds, respectively. Tables 2.1 to 2.4 summarize the values of $\Delta_f H^\theta_{add,DHC}$, $\Delta_f H^\theta_{nonadd, DHC}$, $\Delta_f H^\theta_{add,IHC}$ and $\Delta_f H^\theta_{nonadd, IHC}$ for various different functional groups and molecular fragments.

Example 2.6. Glycerol-1,2-dinitrate has the following molecular structure:

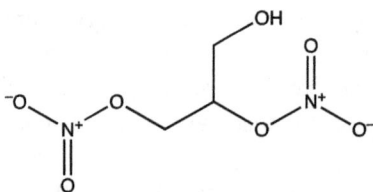

Applying equation (2.13) gives the condensed phase heat of formation as

$$\Delta_f H^\theta(c) = -111.4 + 33.11a - 28.84b + 86.80c - 62.03d$$
$$- 79.25(\Delta_f H^\theta_{add,DHC} + \Delta_f H^\theta_{nonadd,DHC})$$
$$+ 153.3(\Delta_f H^\theta_{add,IHC} + \Delta_f H^\theta_{nonadd,IHC})$$
$$= -111.4 + 33.11(3) - 28.84(6) + 86.80(2) - 62.03(7)$$
$$- 79.25(1 + 0) + 153.3(0 + 0.1)$$
$$= -509.6 \text{ kJ/mol}.$$

The measured $\Delta_f H^\theta(c)$ of this compound is -472.4 kJ/mol [142].

2.2 Energetic compounds with high nitrogen contents

Significant amounts of solid carbon and nonoxidized organic species are produced during the detonation of common high explosives because of the negative oxygen balance of these compounds. However, the energy that nitrogen-rich explosives liberate is due to their high positive heats of formation rather than as a result of the oxidation of the carbon backbone [145]. The main detonation product of materials with high nitrogen contents is N_2 gas, which means that the detonation process is clean [146]. High-nitrogen materials have high densities and good oxygen balances because they

Table 2.1: Summary of the correcting function $\Delta_f H^\theta_{add,DHC}$.

Molecular moieties	Compound	$\Delta_f H^\theta_{add,DHC}$	Example	Exception
Hydroxyl group	Ar–OH	1.2	2 × 1.2 = 2.4 for 4,6-dinitroresorcinol	
	R–OH	1.0	3 × 1.0 = 3.0 for 2-(hydroxymethyl)-2-nitro-1,3-propanediol	
–NH$_2$ or >NH group	More than one –NH$_2$ or C–NH– group	0.6	3 × 0.6 = 1.8 for 2,4,6-trinitrobenzene-1,3,5-triamine	The value of $\Delta_f H^\theta_{add,DHC}$ = 0.7 and 1.4 for the presence of Ar–NH–Ar and 1,2,4-triazole molecular fragments, respectively.
Specific polar groups	–COOH (or –ONH$_4$, –NHCOO–, –NHCONH–, –NHCOC– and –COCO–)	2.8	2.8 for 3,5-dinitrobenzoic acid	
	–COOCO–	2.5	2.5 for bis(1-oxobutyl) peroxide	
	–NHC(=NH)NH–	1.7	1.7 for nitroguanidine (NQ)	
	–COO–C	1.4	1.4 for butanoic acid, 4,4,4-trinitro-, 2,2,2-trinitroethyl ester	
	Cyclic or acyclic ketone	1.2	1.2 for tetramethylolcyclopentanone tetranitrate	
Alkyl or aryl halide	Fluorine atoms attached to non-aromatic carbon atoms	$6n_{CF_2} - 2$, where n_{CF_2} is the number of CF$_2$ groups	6 × 2 – 2 = 10 for 1,1,2,2-tetrafluoro-1,2-dinitroethane	
Nitrate salts	–CH$_n$–NH$_2$ (or –(CH$_n$)$_2$–NH, cyclic –(CH$_n$)$_3$–N, –NH$_n$–NH$_2$ and NH$_4^+$)	1.5	2 × 1.5 = 3.0 for ethylenediamine dinitrate (EDD)	
	–C(=O or NH)–NH$_2$	4.2	4.2 for guanidine nitrate	

Table 2.2: Summary of the correcting function $\Delta_f H^\theta_{\text{nonadd, DHC}}$.

Molecular moieties	Compound	$\Delta_f H^\theta_{\text{nonadd, DHC}}$	Example	Exception
Ether	Acyclic ether	0.5	2,2'-[Oxybis(methylene)]bis[2-[(nitrooxy)methyl]propane-1,3-diyl] tetranitrate	Ar–O–Ar and Ar–O–R where R is an alkyl group with more than one carbon atom
	Six member cyclic ether	1.7	D-glucopyranose pentanitrate	
	–O–C–O–C–O–	3.4	Saccharose octanitrate	
Alkyl fluoride	Monofluoro derivative of R–F	2	Fluorotrinitromethane	

Table 2.3: Summary of the correcting function $\Delta_f H^\theta_{\text{add, IHC}}$.

Molecular moieties	Compound	$\Delta_f H^\theta_{\text{add, IHC}}$	Example	Exception
Molecular fragments –N=N– (or –N–N–N– and –N–N–N–N–), >N–O⁻ (or azido group)	–N=N– (or –N–N–N– and –N–N–N–N–)	0.6	$2 \times 0.6 = 1.2$ for 2,6-bis(picrylazo)-3,5-dinitropyridine	
	$\overset{+}{\text{N}}$–O⁻	0.7	$3 \times 0.7 = 2.1$ for benzo[1,2-c:3,4-c':5,6-c'']tris[1,2,5]oxadiazole, 1,4,7-trioxide (benzenetrifuroxan)	
Nitro groups attached to a non-aromatic four-membered ring	Attachment of –NO$_2$ groups to a non-aromatic four-membered ring	0.45	$8 \times 0.45 = 3.6$ for octanitrocubane	
Number of carbocyclic aromatic rings (n_{Ar})	For energetic compounds containing only carbocyclic aromatic rings	$(n_{\text{Ar}} - 1) \times 0.3$	$(2 - 1) \times 0.3 = 0.3$ for 2,2',4,4',6,6'-hexanitrobiphenyl	

Table 2.4: Summary of the correcting function $\Delta_f H^\theta_{\text{nonadd, IHC}}$.

Molecular moieties	Compound	$\Delta_f H^\theta_{\text{nonadd, IHC}}$	Example	Exception
Cyclic and acyclic nitramine (or nitroso amine) functional groups	Acyclic nitramine	0.8	N-Methyl-N-nitro-methanamine	
	Cyclic nitramine (or nitroso amine)	0.5	1,4-Dinitropiperazine	
$R-NO_2$, $R-ONO_2$, $R-CN$, mono and polynitro-benzene, $-NH-NO_2$ (or $-NF-NO_2$), $-ONO$ (nitrite), $-O-OH$ and azide derivatives of carbocyclic aromatic compounds or $R-N_3$	Attachment of more than one $-NO_2$ group to non-aromatic carbon atom(s)	0.6	1,1-Dinitropropane	(a) for the presence $-C(NO_2)_3$, $\Delta_f H^\theta_{\text{nonadd, IHC}} = 0.8$ (e. g. 2,2,2-trinitroethanol); (b) for linear mono-nitroalkanes, $\Delta_f H^\theta_{\text{nonadd, IHC}} = 0.6$ (e. g. 1-nitrobutane); (c) for hydrogen-free nitro-alkanes, $\Delta_f H^\theta_{\text{nonadd, IHC}} = 2.2$ (e. g. for tetranitromethane)
	Attachment of the $-N_3$ group to a carbocyclic aromatic ring (or $R-N_3$)	0.7	(Azidomethyl)-benzene	
	$R-ONO_2$ (other functional groups except the $-NO_2$ group are absent)	0.5	Ethylene glycol, dinitrate (nitroglycol)	
	$R-ONO_2$ (in the presence of the other functional groups)	0.1	N-Butyl-N-(2-nitroxyethyl)nitramine (BuNENA)	
	Mono and polynitrobenzene	0.7	1,2-dinitrobenzene	
	$-ONO$ (nitrite)	0.5	Propyl nitrite	
	$-O-OH$	0.5	tert-Butyl hydroperoxide	
	$-CN$	0.4	Trinitroacetonitrile	
	$-NH-NO_2$ (or $-NF-NO_2$)	0.2	Ethylenedinitramine (EDNA)	
Furan derivatives and cyclic ethers smaller than a six membered ring	Furan derivatives	0.25	2-Nitrofuran	
	Cyclic ethers amller than a six membered ring	0.5	α-Epoxyconduritole tetranitrate	

Table 2.4 (continued)

Molecular moieties	Compound	$\Delta_f H^\theta_{\text{nonadd, IHC}}$	Example	Exception
Attachment of alkyl groups to aromatic ring	Attachment of methyl groups to an aromatic ring	0.5	2,4,6-Trinitrotoluene (TNT)	
	Attachment of longer carbon-chain alkyl groups (or –CH=CH–) to an aromatic ring	0.6	1-Ethyl-2-nitrobenzene	
Cyclic and acyclic peroxide	Cyclic and acyclic peroxide with general formula R_1–O–O–R_2 in which R_1 or R_2 has four or less carbon atoms	0.7	Hexamethylenetriperoxide Diamine (HMTD)	
Alkyl or aryl halide	Monochloro derivative of Ar–Cl	0.2	1-Chloro-2-nitrobenzene	
	Monochloro derivative of R–Cl	0.5	1-Chloro-1,1-dinitroethane	
	Monobromo derivative of R–Br	0.8	Bromotrinitromethane	
	Monoiodo derivative of R–I	1.2	Iodotrinitromethane	

have a low carbon and hydrogen content [147]. Common nitramine high explosives have nitrogen contents of up to 38 % (in RDX, HMX and Cl-20), while some derivatives of triazole, tetrazole, triazine, tetrazine, furazan and organic nitrogen-containing chains have up to 88 % nitrogen-contents. High-nitrogen materials have been defined as those containing more than 50 % w/w nitrogen in their molecular structure [148]. High-nitrogen compounds are suitable candidates for insensitive high explosives [149], gun propellants [150, 151], and clean gas generators in vehicle airbags [152]. Two different empirical methods are introduced here for the calculation of the condensed phase heat of formation of these compounds.

2.2.1 Using the molecular structure

It has been shown that the molecular structure of high-nitrogen materials can be used to predict their condensed phase heat of formation as follows [153]:

$$\Delta_f H^\theta(c) = 39.24a - 40.01b + 83.63c - 49.61d + 115.5 \sum_i IF_i - 177.4 \sum_j DF_j, \quad (2.14)$$

where IF and DF are correcting factors, which can be estimated as follows.
(1) Azido group: For the attachment of the $-N_3$ group to a tetrazole or tetrazine ring, as well as to aliphatic N-containing compounds, IF is the number of azido groups. Meanwhile, the contribution of $-N_3$ for high-N compounds containing only triazole and triazine rings is zero.
(2) Azo and azoxy groups: In tetrazine, triazine, triazole and furazan derivatives which contain $-N{=}N-$ or $-N{=}N^+O^--$ bridges, $IF = 1.5$. For tetrazole derivatives, $IF = 0.2$.
(3) Guanidino, carbonyl and hydroxide groups: if any of these functional groups are present, $DF = 1.2 \times$ the number of these groups in the molecule.
(4) Amino groups: $DF(-NH_2) = 0.6 \times$ (the number of $-NH_2$ groups) and $DF(-NH-) = 0.3 \times$ (the number of $-NH-$ groups).

Example 2.7. The condensed phase heat of formation of 3,6-diazido-1,2,4,5-tetrazine with the following molecular structure is calculated as

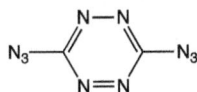

$$\Delta_f H^\theta(c) = 39.24a - 40.01b + 83.63c - 49.61d + 115.5 \sum_i IF_i - 177.4 \sum_j DF_j$$

$$= 39.24(2) - 40.01(0) + 83.63(10) - 49.61(0) + 115.5(2) - 177.4(0)$$

$$= 1146 \, \text{kJ/mol}.$$

The experimental $\Delta_f H^\theta(c)$ was reported to be 1101 kJ/mol [154].

2.2.2 Gas phase information

It was shown that the value of $\Delta_f H^\theta(c)$ can be calculated using the $\Delta_f H^\theta(g)$ value obtained from either the B3LYP/6-31G* or PM6 method as follows [134]:

$$\Delta_f H^\theta(c) = -57.28 + 0.9350[\Delta_f H^\theta(g)]_{\text{B3LYP/6-31G}*}$$
$$+ 27.86 \sum_i ICF_i - 30.06 \sum_j DCF_j \tag{2.15}$$

$$\Delta_f H^\theta(c) = -34.99 + 0.8692[\Delta_f H^\theta(g)]_{\text{PM6}}$$
$$+ 30.64 \sum_i ICF_i - 25.95 \sum_j DCF_j, \tag{2.16}$$

where *ICF* and *DCF* are increasing and decreasing correcting factors, respectively, which are estimated according to the following.
(1) Azido group: The value of *ICF* equals two times the number of azido groups in the molecule.
(2) Azo and azoxy groups: The value of *ICF* equals the total number of −N=N− or −N=N⁺O⁻ bridges.
(3) Hydrogen bonding effect: The value of *DCF* is the number of hydrogen bonds in each molecule.
(4) Ethylene groups: The value of *DCF* is two times the total number of ethylene groups in the molecule.

Appendix B shows the calculated $[\Delta_f H^\theta(g)]_{\text{B3LYP/6-31G}*}$ and $[\Delta_f H^\theta(g)]_{\text{PM6}}$. Table B.1 provides the calculated total energies and formation enthalpies (at 0 K) for 100 energetic materials with high nitrogen contents by the B3LYP/6-31G*. Table B.2 also gives the predicted gas phase standard enthalpies of formation for 100 energetic materials with high nitrogen content.

Example 2.8. The calculated gas phase heat of formation of 1,2-di(1H-tetrazol-5-yl)ethane with the following molecular structure is $[\Delta_f H^\theta(g)]_{\text{B3LYP/6-31G}*}$ = 639.5 kJ/mol and $[\Delta_f H^\theta(g)]_{\text{PM6}}$ = 613.9 kJ/mol:

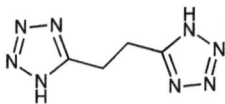

The use of equations (2.15) and (2.16) gives the values of $\Delta_f H^\theta(c)$ as

$$\Delta_f H^\theta(c) = -57.28 + 0.9350[\Delta_f H^\theta(g)]_{\text{B3LYP/6-31G}*} + 27.86 \sum_i ICF_i - 30.06 \sum_j DCF_j$$
$$= -57.28 + 0.9350(639.5)_{\text{B3LYP}} + 27.86(0) - 30.06(2 + 2)$$
$$= 420.4 \text{ kJ/mol,}$$

$$\Delta_f H^\theta(c) = -34.99 + 0.8692[\Delta_f H^\theta(g)]_{PM6} + 30.64 \sum_i ICF_i - 25.95 \sum_j DCF_j$$

$$= -34.99 + 0.8692(613.9)_{PM6} + 30.64(0) - 25.95(2 + 2)$$

$$= 398.4 \, \text{kJ/mol.}$$

The experimental $\Delta_f H^\theta(c)$ value was reported to be 444.4 kJ/mol [155].

2.3 The condensed phase heat of formation of energetic ionic liquids and salts

2.3.1 Complex approach

The presence of energetic cations or anions in some classes of ionic liquids or salts provides energetic ionic liquids or salts. Energetic ionic liquids or salts (EILoS) can contain high nitrogen (high-N) organic cations and bulky anions. Anions include one or more energetic groups, e. g. $-NO_2$, $-N_3$, and $-CN$. Since EILoS can give suitable thermally stability, they may be used as explosives, pyrotechnics, and propellants [159–161]. The value of the condensed phase heat of formation of an EILoS is very important for the assessment of its detonation velocity and detonation pressure [162]. For an EILoS with the general formula $(cation)_i(anion)_j$, computation of its $\Delta_f H^\theta(c)$ can be done using gas phase heats of formation of anion and cation as well as the heat of phase transition according to Hess's law of constant summation (Born–Haber energy cycle) as [163]:

$$\Delta_f H^\theta(c)[EILoS] = \sum_i \Delta_f H_i^\theta(g)[cation] + \sum_j \Delta_f H_j^\theta(g)[anion] - \Delta H_{PT} \qquad (2.17)$$

where $\Delta_f H^\theta(c)[EILoS]$ is the condensed phase heat of formation of EILoS; $\sum_i \Delta_f H_i^\theta(g)$ [cation] and $\sum_j \Delta_f H_j^\theta(g)[anion]$ are the sum of gas phase heats of formation of cations and anions, respectively; ΔH_{PT} is the heat of phase transition. Gao et al. [164] used equation (2.17) to compute $\Delta_f H^\theta(c)[EILoS]$ for 119 energetic salts including imidazolium, triazolium, and tetrazolium-based cations. They also used isodesmic reactions to compute $\sum_i \Delta_f H_i^\theta(g)[cation]$ and $\sum_j \Delta_f H_j^\theta(g)[anion]$. For computation of ΔH_{PT}, they used the method of Jenkins et al. [165]. They considered all species as closed shells in nature. They carried out various geometric optimization and frequency analyses for molecules with more than 10 heavy atoms using B3-LYP functional with 6-31+G** basis set. They calculated single energy points at the MP2-(full)/6-311++G** level. They characterized all of the optimized structures to obtain true local energy minima on the potential energy surface without imaginary frequencies. They computed the molecular energies (proton affinities and ionization energies) at the G215 or G2MP2 level. Zhang et al. [166] studied the effects of different energetic substituents containing $-NO_2$, $-NF_2$, $-CN$, $-N_3$, and $-NH_2$ $\Delta_f H^\theta(c)[EILoS]$ for some tetrazole salts using a similar complex approach as that used by Gao et al. [164]. They indicated that the presence of energetic groups in most cases can increase $\Delta_f H^\theta(c)[EILoS]$. The use of this approach has several shortcomings because it requires a high-speed computer, specific computer codes, and expert users. Thus, two simple QSPR methods based on fragments of EILoS for prediction of $\Delta_f H^\theta(c)[EILoS]$ are discussed here [167, 168].

2.3.2 QSPR methods

2.3.2.1 Imidazolium-based ionic liquids or salts

A simple approach was introduced to estimate $\Delta_f H^\theta(c)$ of imidazolium-based ionic liquids or salts [167]. For some imidazolium-based ionic liquids or salts, the reliability of the model was higher than the available outputs of complex quantum mechanical methods [167]. This approach is based on the following form [167]:

$$\Delta_f H^\theta(c)[IMILoS] = 1594 - 76.32b_{cat} - 433.0e_{ani} + 289.3\Delta_f H^\theta_{Inc}[IMILoS]$$
$$- 304.3\Delta_f H^\theta_{Dec}[IMILoS] \tag{2.18}$$

where $\Delta_f H^\theta(c)[IMILoS]$ is the condensed phase heat of formation of imidazolium-based ionic liquids or salts; b_{cat} and e_{ani} denote the number of hydrogen and fluorine atoms in cation and anion, respectively; $\Delta_f H^\theta_{Inc}[IMILoS]$ and $\Delta_f H^\theta_{Dec}[IMILoS]$ contribute to increasing and decreasing heat contents in imidazolium-based ionic liquids or salts, respectively [167]. Table 2.5 summarizes the values of $\Delta_f H^\theta_{Inc}[IMILoS]$ and $\Delta_f H^\theta_{Dec}[IMILoS]$.

Since the coefficients of b_{cat} and e_{ani} have negative signs, decreasing their values gives a more positive value of $\Delta_f H^\theta(c)[IMILoS]$. Thus, it is desirable to design imidazolium-based ionic liquids or salts with low values of b_{cat} and e_{ani} because they provide high detonation or combustion performance. Moreover, the existence and absence of $\Delta_f H^\theta_{Inc}[IMILoS]$ and $\Delta_f H^\theta_{Dec}[IMILoS]$, respectively, can help to improve the stored chemical energy in a designed sample.

Example 2.9. Calculate the value of $\Delta_f H^\theta(c)[IMILoS]$ for the following ionic liquid:

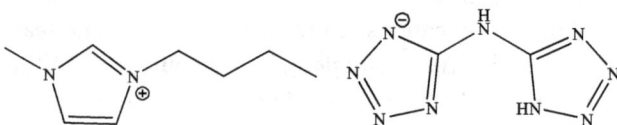

The use of equation (2.18) and Table 2.5 provides the value of $\Delta_f H^\theta(c)[IMILoS]$ as:

$$\Delta_f H^\theta(c)[IMILoS] = 1594 - 76.32b_{cat} - 433.0e_{ani} + 289.3\Delta_f H^\theta_{Inc}[IMILoS]$$
$$- 304.3\Delta_f H^\theta_{Dec}[IMILoS]$$
$$= 1594 - 76.32 \times 30 - 433.0 \times 0 + 289.3 \times 0.4(3-1) - 304.3 \times 0$$
$$= 680 \text{ kJ/mol}$$

The reported value of $\Delta_f H^\theta(c)[IMILoS]$ is 682.3 [169].

Table 2.5: Different values of $\Delta_f H^\theta_{Inc}[IMILoS]$ and $\Delta_f H^\theta_{Dec}[IMILoS]$.

Cation	Anion	$\Delta_f H^\theta_{Inc}[IMILoS]$	$\Delta_f H^\theta_{Dec}[IMILoS]$	Condition
imidazolium–(CH$_2$)$_n$	bis(tetrazolyl)amine anion	$0.4(n-1)$	0	$1 \le n \le 7$
	dicyanamide / azotetrazolate anion	0	$1.75 - 0.33n$	$1 \le n \le 3$
	ethyl sulfate	0	4.4	$n = 1$
	bis(trifluoromethylsulfonyl)imide (NTf$_2$)	$3.5 - \frac{9-n}{2}$	0	$n \le 9$
		0	4.0	$n \ge 13$
	hexafluorophosphate (PF$_6$)	$3.0 - \frac{9-n}{2}$	0	$3 \le n \le 9$
		0	4.3	$n \ge 11$
	tetrafluoroborate (BF$_4$)	0	1.2	$n = 3$
		0	5.7	$n \ge 11$
dimethylimidazolium	bis(tetrazolyl)amine anion	0	0.3	–
trimethylimidazolium	azotetrazolate anion	3.0	0	–
nitro-substituted imidazolium (O$_2$N–)	azotetrazolate anion			

Table 2.5 (continued)

Cation	Anion	$\Delta_f H^\theta_{Inc}[IMILoS]$	$\Delta_f H^\theta_{Dec}[IMILoS]$	Condition
		0	$-0.4n + 2.9$	$3 \leq n \leq 4$
	Br$^\ominus$			
		0	0.8	$n \leq 3$
		3.2	0	$n \geq 7$

2.3.2.2 Triazolium-based energetic ionic liquids or salts

A simple method has been introduced to estimate $\Delta_f H^\theta(c)$ of triazolium-based energetic ionic liquids or salts [168]. It requires some specific elemental composition of cations and anions as well as two correcting functions. It is given as follows [168]:

$$\Delta_f H^\theta(c)[TAILoS] = -27.31b_{cat} + 102.7c_{cat} + 259.8a_{ani} - 319.2b_{ani} + 45.32c_{ani} - 125.9d_{ani}$$
$$+ 632.6f_{ani} + 79.78\Delta_f H^\theta_{Inc}[TAILoS] - 74.50\Delta_f H^\theta_{Dec}[TAILoS] \quad (2.19)$$

where $\Delta_f H^\theta(c)[TAILoS]$ is the value of $\Delta_f H^\theta(c)$ for triazolium-based energetic ionic liquids or salts; c_{cat} is the number of nitrogen atoms in cation; a_{ani}, b_{ani}, c_{ani}, d_{ani} and f_{ani} give the number of carbon, hydrogen, nitrogen, oxygen, and chlorine atoms in the anion, respectively; $\Delta_f H^\theta_{Inc}[TAILoS]$ and $\Delta_f H^\theta_{Dec}[TAILoS]$ give increasing and decreasing functions in the triazolium-based EILoS, respectively. Table 2.6 gives the summary of the values of $\Delta_f H^\theta_{Inc}[TAILoS]$ and $\Delta_f H^\theta_{Dec}[TAILoS]$.

Example 2.10. Predict the value of $\Delta_f H^\theta(c)[TAILoS]$ for the following ionic liquid:

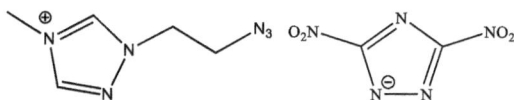

The use of equation (2.19) and Table 2.6 gives the value of $\Delta_f H^\theta(c)[TAILoS]$ as:

$$\Delta_f H^\theta(c)[TAILoS] = -27.31b_{cat} + 102.7c_{cat} + 259.8a_{ani} - 319.2b_{ani} + 45.32c_{ani} - 125.9d_{ani}$$
$$+ 632.6f_{ani} + 79.78\Delta_f H^\theta_{Inc}[TAILoS] - 74.50\Delta_f H^\theta_{Dec}[TAILoS]$$
$$= -27.31 \times 9 + 102.7 \times 6 + 259.8 \times 2 + 45.23 \times 5 - 125.9 \times 4 + 79.78 \times 1.5$$
$$= 732.2 \text{ kJ/mol}$$

The measured value of $\Delta_f H^\theta(c)[TAILoS]$ is 771.3 [170]. The calculated value of $\Delta_f H^\theta(c)[TAILoS]$ by complex quantum mechanical method using equation (2.17) is 576.0 kJ/mol [164], which has a larger deviation (-195.3 kJ/mol).

Table 2.6: Contribution of structural parameters in predicting $\Delta_f H^\theta_{Inc}[TAILoS]$ and $\Delta_f H^\theta_{Dec}[TAILoS]$.

Cation	Anion	$\Delta_f H^\theta_{Inc}[TAILoS]$	$\Delta_f H^\theta_{Dec}[TAILoS]$
	NO_3^-	0	1.5
		2.0	0
	ClO_4^-	0	1.5
		2.5	0
		0	2.0
		2.5	0
		5.0	0
		0	2.5
		0	3.5
		1.5	0
		1.5	0
		0	1.5
		0	3.0

Table 2.6 (continued)

Cation	Anion	$\Delta_f H^\theta_{Inc}[TAILoS]$	$\Delta_f H^\theta_{Dec}[TAILoS]$
		0	5.5
		3.0	0
		1.5	0
		0	3.5

Appendix C including Tables C.1 and C.2 shows the measured and calculated values of $\Delta_f H^\theta(c)[IMILoS]$ and $\Delta_f H^\theta(c)[TAILoS]$ for some energetic ionic liquids and salts.

2.4 Summary

This chapter has introduced different empirical methods for the prediction of the condensed phase heat of formation of important classes of energetic compounds, as well as of high-nitrogen compounds (i. e. compounds with more than 50 % w/w nitrogen in their molecular structure). Equations (2.6) and (2.10) provide the simplest approaches for the prediction of the $\Delta_f H^\theta(c)$ values of nitroaromatics, and also for nitramines, nitrate esters, nitroaliphatics and related energetic compounds. The other correlations outlined in Section 2.1.2 give more reliable predictions, but they are considerably more complex. Among the correlations which were introduced, equation (2.13) provides the best method for the reliable calculation of $\Delta_f H^\theta(c)$ for different types of energetic compounds containing different types of energetic groups, i. e. $-NO_2$, $-ONO_2$, $-NNO_2$, $-ON=O$, $-O-O-$ and $-N_3$. For high-nitrogen compounds, equation (2.15) is simpler than equation (2.16). The reliability of both correlations is good, but equation (2.16) should be used preferably for high-nitrogen compounds containing unfamiliar molecular fragments.

Section 2.3 introduces equation (2.17) as a complex quantum mechanical approach for the calculation of $\Delta_f H^\theta(c)[EILoS]$. Equations (2.18) and (2.19) give two simple and reliable correlations for predicting the values of $\Delta_f H^\theta(c)[IMILoS]$ and $\Delta_f H^\theta(c)[TAILoS]$, respectively.

3 Melting point

The knowledge of the melting point of the desired compound can be used for its identification and purification. It may be used to calculate other properties, e. g. aqueous solubility, liquid viscosity, and vapor pressure [171]. Since there is a close relationship of melting point to solubility, it is important in environmental studies and the assessment of the purity of a chemical substance [172]. In the measurement of melting points some difficulties may occur such as the relationship of the melting point to the phase transition from the solid to the liquid state and the possible existence of different crystalline structures below the melting point [173]. The melting point of organic compounds containing energetic groups is a significant property especially for melting cast explosives [174]. It is important to have reliable methods for the prediction of melting points of different classes of organic energetic compounds because they are significant in the design of low-melting-point candidates for casting medium.

The search for new energetic compounds with "ideal" physical properties is a problem of the utmost importance to both research chemists and chemical industry. Since the melting point is one of the fundamental physical properties which is used in the identification and purification of a chemical, and is also highly valuable for the calculation of other important physicochemical properties such as the vapor pressure and aqueous solubility, it is important to have reliable methods which can be used to predict the melting point of an energetic compound. Group additivity, QSPR, quantum mechanical and empirical methods are all different approaches which have been developed to enable prediction of the melting points of different classes of organic compounds and ionic liquids. Some simple and reliable empirical methods are discussed in detail in this chapter, and the other methods are also briefly described.

https://doi.org/10.1515/9783110740158-003

3.1 Group additivity, QSPR and quantum mechanical methods

Among the different approaches which can be used to predict the melting points of organic compounds, group additivity methods are widely used [130]. The group additivity methods can predict the melting point of a desired organic compound by summing the number of each group multiplied by its contribution [130]. Although they are simple and provide quick estimations, many have questionable accuracy and their reliability with respect to the prediction of melting points of organic energetic compounds is unknown. The Joback–Reid [175] approach for the estimation of the melting points of pure compounds is the simplest group additivity method. The estimated melting points (T_m) from the Joback–Reid group additivity method [175] can be estimated as follows:

$$T_m = 122.5 + \sum n_i \text{GAV}_i, \tag{3.1}$$

where T_m is in K; n_i is the number of groups of type i and GAV_i is the group contribution of the melting points of group i in the molecule. It was found that the results predicted using the group-contribution method of Joback–Reid [175] show an average deviation of 37.6 % for 60 carbocyclic nitroaromatic compounds which were considered [176]. Thus, the group contribution method can only be used to obtain a very approximate guess. Some of the other group additivity methods include those of Lydersen [177], Ambrose [178], Klincewicz and Reid [179], Lyman et al. [180], Constantinou et al. [181–184], Marrero-Morejón and Pardillo-Fontdevilla [185], and Marrero and Gani [186]. In contrast to other physicochemical properties, melting points are not well estimated by the group additivity methods [177, 187–189].

Quantum mechanical calculations are another approach that has been used for simulating solid to liquid phase transitions in energetic materials, in order to predict their melting points [190–193]. Molecular dynamics is a complex method which can be used to simulate solid to liquid phase transitions in energetic materials to predict their melting points [190–193]. Due to the free energy barrier for the formation of a liquid-solid interface, estimation of the melting point can be considered difficult problems through molecular dynamics. This situation can cause superheating in a perfect crystal, which results in an overestimation of the melting point. Various ways of performing the simulations have been investigated in order to determine the most practical one for use for energetic materials [190–193]. Methods of QSPR have been developed for organic molecules – including a number of drugs and/or homologous series [193–196].

PARC (Sparc Performs Automated Reasoning in Chemistry) is a complex physicochemical calculator, which can be used to estimate chemical reactivity parameters and physical properties of organic molecules [197]. It uses mechanisms to calculate these properties by incorporating trained parameters derived from experimental measurements and best fit using linear regression [197]. Whiteside et al. [198] used the SPARC platform to estimate the entropy of fusion, enthalpy of fusion, and the melting point of organic compounds through three models. The first model combines interaction terms and physical descriptors for the calculation of the entropy of fusion. The second model is a function of the entropy of fusion, boiling point, and flexibility of the molecule to estimate the enthalpy of fusion. The third model is based on the division of the enthalpy of fusion by the entropy of fusion for the prediction of the melting point. The predicted root-mean-square (RMS) deviations of 904 organic compounds for the entropy, enthalpy, and melting models are 12.5 J mol^{-1} K^{-1}, 4.87 kJ mol^{-1}, and 54.4 K, respectively.

Yalkowsky and coworkers [199–202] have developed UPPER (Unified Physical Property Estimation Relationships) as a system of empirical and theoretical relationships for the estimation of physicochemical properties of organic molecules. The UPPER can calculate the entropy of fusion, enthalpy of fusion, and melting point of organic compounds. It has been used for a wide range of organic compounds (over 2000 compounds) as compared to SPARC with a similar pattern for calculation of melting point, i. e. the ratio of enthalpy of fusion to the entropy of fusion. It uses the group additivity method for the calculation of the enthalpy of fusion. It also uses molecular geometry for the assessment of the degree of restriction of molecular motion in the crystal to that of the liquid, i. e. symmetry, eccentricity, chirality, flexibility, and hydrogen bonding.

Al-Fakih et al. [172] introduced a QSPR model with four complex descriptors for the prediction of melting points of 92 carbocyclic nitroaromatic compounds based on the proposed penalized adaptive bridge (PBridge) as:

$$T_m = 11.034 + 0.542 Eig02\_AEA(dm) - 42.311 Mor29v - 2.547 Mor13u + 4.592 D/Dtr09,$$
$$(3.2)$$

where $Eig02\_AEA(dm)$ is edge adjacency indiex corresponding to Eigenvalue n. 2 from edge adjacency mat. weighted by dipole moment; $Mor29v$ and $Mor13u$ are two 3D-MoRSE descriptors corresponding to signal 29/weighted by van der Waals volume and signal 13/unweighted, respectively; $D/Dtr09$ is ring descriptors corresponding to distance/detour ring index of order 9.

3.2 Simple empirical methods on the basis of molecular structure

Several simple correlations have recently been introduced to predict the melting points of certain classes of energetic compounds. If impurities are present in the energetic compounds, or if the energetic compound exhibits thermal instability, experimental determination of the melting point may be thwarted. It is important to select a reliable predictive model for organic compounds containing energetic functional groups such as Ar–NO$_2$, C–NO$_2$, C–ONO$_2$ or N–NO$_2$. According to Carnelley's rule [203, 204], the more symmetrical the organic isomer is, the higher its melting point. The dipole moment is one of the factors which directly controls the melting point. Thus, the sum of all of the local dipole moments has a more pronounced effect than the net dipole moment on the melting point. Due to interactions between local dipole moments of neighboring atoms or groups, molecular interactions can result in close proximity of the molecules in the crystal. It can be expected that the more symmetrical paraisomers and the local dipole moments of some polar groups have a distinct effect in increasing the melting point. The presence of alkyl and alkoxy groups attached to nitroaromatic rings can decrease the planarity of the molecules, which in turn reduces the packing efficiency of molecules in the crystals. This results in a decrease in the interaction between local dipole moments of neighboring nitro groups. Although the attachment of a polar nitramine group may increase the molecular interactions, the presence of an alkyl substituent can change its effect because it can make the molecule less planar. The molecules become further apart if an alkyl nitramine group is introduced. In this section, some of the best available, simple methods for predicting melting points for several selected classes of energetic compounds will be reviewed.

3.2.1 Nitroaromatic compounds

The study of various carbocyclic nitroaromatic organic compounds with general formula $C_aH_bN_cO_d$ has shown that the following equation can be used to predict the melting point [176]:

$$T_m = 282.96 - 2.7543b + 46.570c + 94.318T_{SFG} + 54.752T_{o,p}, \tag{3.3}$$

where T_{SFG} is the contribution of a specific functional group and $T_{o,p}$ is a parameter that can be applied to disubstituted benzene rings. The values of T_{SFG} and $T_{o,p}$ are predicted based on the following conditions:

(1) T_{SFG}: If −OH and −NH$_2$ are *ortho* to the −NO$_2$ group, the value of T_{SFG} equals zero, whereas if the −NH$_2$ and −OH groups in are *meta* or *para* positions relative to the −NO$_2$ group then T_{SFG} has the value 0.3. T_{SFG} = 1.0 if the −COOH, −CON− or −COO− functional groups are present. T_{SFG} also has the value −1.2 if the nitramine (N−NO$_2$) functional group is present.

(2) $T_{o,p}$: The value of $T_{o,p}$ = 1.0 for *para* disubstituted benzene rings. The presence of alkyl (−R) or alkoxy (−OR) groups in *ortho* positions relative to the −NO$_2$ group may result in a decrease in the melting point and therefore $T_{o,p}$ is equal to −0.7.

Example 3.1. 2,6-Bis(picryamino)-3,5-dinitropyridine (PYX) has the following molecular structure:

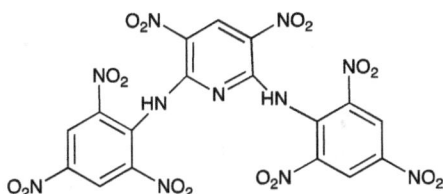

The use of equation (3.3) gives T_m as

$$T_m = 282.96 - 2.7543b + 46.570c + 94.318T_{SFG} + 54.752T_{o,p}$$
$$= 282.96 - 2.7543(7) + 46.570(11) + 94.318(0) + 54.752(0)$$
$$= 775.95\,K.$$

The experimental T_m was reported to be 733 K [142].

3.2.2 Polynitro arene and polynitro heteroarene compounds

Another correlation has recently been introduced that can be applied to a wider classes of energetic compounds, including polynitro arenes and polynitro hetero-arenes, and is as follows [205]:

$$T_m = 355.0 + 3.33a - 5.63b + 14.57c + 90.83ISSP - 63.75DSSP, \tag{3.4}$$

where *ISSP* and *DSSP* are increasing and decreasing effects of some specific structural features, which can be specified according to the following situations.

ISSP
(1) The presence of $-NH_2$ groups:
   (a) Polynitro arenes: The value of *ISSP* is 0.5 if $n_{NH_2}/n_{NO_2} < 0.5$ except for monon-itro anilines, in which *ISSP* is equal to 0.5 and 0.0 for *para* nitro aniline and *ortho* (or *meta*) aniline, respectively. If $n_{NH_2}/n_{NO_2} \geq 0.5$, *ISSP* equals 2.0 for the attachment of $-NH_2$ groups to carbocyclic nitroaromatic compounds.
   (b) Polynitro heteroarenes: If amino groups are attached to heterocyclic aromatic compounds, *ISSP* = 1.0.
   Since the participation of a functional group in intramolecular hydrogen bonding reduces its ability to form intermolecular hydrogen bonds, the presence of n_{NO_2} in *ortho* positions or in positions close to n_{NH_2} groups has no appreciable effect, e. g. in *o*-nitroaniline.
(2) The presence of certain functional groups: The value of *ISSP* = 1.0 if one $-COOH$ group or two $-OH$ functional groups are attached to an aromatic ring.

(3) The existence of specific structural parameters: If the [structure] or –C=C– groups are attached to an aromatic ring, *ISSP* equals 1.0.

(4) The presence of some specific groups in disubstituted benzene rings (in para position with respect to an $-NO_2$ group): If the $-NO_2$, R_2N- or $-N-C(=O)-$ groups are located in the *para* position with respect to the nitro group, *ISSP* equals 1.0.

DSSP

(1) The attachment of only alkyl or alkoxy groups to nitroaromatic rings: If $n_{NO_2}/n_{R,OR} \leq 1/3$ (where $n_{R,OR}$ is the number of $-R$ or $-OR$ groups attached to a nitroaromatic ring), then *DSSP* = 0.0. For $1/3 < n_{NO_2}/n_{R,OR} \leq 1$, the value of *DSSP* is equal to 1.0. The *DSSP* value is also 0.5 for $n_{NO_2}/n_{R,OR} > 1$.

(2) The presence of alkyl nitramine groups attached to aromatic rings: For the presence of $R-N-NO_2$, *DSSP* $= 0.7 \times n_{RNNO_2}$, where n_{RNNO_2} is the number of alkyl nitramine groups.

(3) The presence of specific structural parameters: For the [structure] N^+-O^- (or [structure]) and [structure] rings, the values of *DSSP* are 1.0 and 2.0, respectively.

(4) Mononitro substituted carbocylic aromatic compounds: The value of *DSSP* is equal to 1.0 only for mononitro substituted carbocylic aromatic compounds.

Different effects of *ISSP* and *DSSP* in polycyclic nitroaromatic compounds

The presence of more than two nitroaromatic rings can result in different situations. If three aromatic rings are present and if $n_{NO_2}/n_{Ar} > 2.5$, then *ISSP* equals 2.0. However, if four aromatic rings are present and if $n_{NO_2}/n_{Ar} > 2.5$, then *DSSP* is 1.0.

Example 3.2. Consider the following molecular structure:

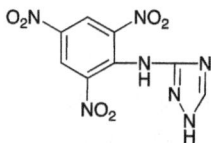

The calculated T_m by equation (3.4) is as follows:

$$T_m \text{ (K)} = 355.0 + 3.33a - 5.63b + 14.57c + 90.83ISSP - 63.75DSSP$$
$$= 355.0 + 3.33(8) - 5.63(5) + 14.57(7) + 90.83(1) - 63.75(0)$$
$$= 546.3 \text{ K.}$$

The measured T_m value was reported to be 583 K [206].

3.2.3 Nitramines, nitrate esters, nitrate salts and nitroaliphatics

The following equation is the simplest approach to predict the melting points of nitramines, nitrate esters, nitrate salts and nitroaliphatics [207]:

$$T_m = 220.47 + 30.220c + 24.780d - 68.691C_{SFG} - 25.891n_{N-NO_2}, \qquad (3.5)$$

where n_{N-NO_2} is the number of nitramine groups in the energetic compound and C_{SFG} is the contribution of specific functional groups, which can be specified as follows:
(1) Nitrate salt: $C_{SFG} = 0$.
(2) Nitrate groups: C_{SFG} has the values 2.0, 2.5 and 3.5 for nitrated energetic compounds, which have one, two, and three or four $-O-NO_2$ groups, respectively.
(3) Hydroxyl group: $C_{SFG} = -1.0$ for nitro or nitrated energetic compounds that have at least one $-OH$ functional group.
(4) $-C(NO_2)_3$ group: C_{SFG} has the values 3.0 and 4.0 for the presence of one and more than one $C(NO_2)_3$ group, respectively.
(5) Mononitro compounds: $C_{SFG} = 1.5$ for mono nitro compounds which have the general formula $R'CH-RCH-NO_2$ (R or R' = $-H$, alkyl).

Example 3.3. For $HOCH_2C(CH_3)_2NO_2$, the predicted melting point using equation (3.5) is calculated as follows:

$$
\begin{aligned}
T_m &= 220.47 + 30.220c + 24.780d - 68.691C_{SFG} - 25.891n_{N-NO_2} \\
&= 220.47 + 30.220(1) + 24.780(3) - 68.691(-1) - 25.891(0) \\
&= 393.72\,\text{K}.
\end{aligned}
$$

The measured T_m value was reported to be 361.7 K [142].

3.2.4 Nonaromatic energetic compounds

It was found that the elemental composition, as well as positive (T^+) and negative (T^-) correcting terms are important factors to include for nonaromatic energetic compounds, since this results in a more reliable correlation than equation (3.5). The new correlation is given as follows [189]:

$$T_m = 281.7 + 28.97a - 12.08b + 29.75c - 9.966d + 102.92T^+ - 110.11T^-. \qquad (3.6)$$

The values of T^+ and T^- depend on some functional groups and structural parameters that are specified as follows.
(1) Nitroaliphatics and nitrate esters:
  (a) $C_nH_{2n+1}(NO_2 \text{ or } ONO_2)_{m=1 \text{ or } 2}$: The values of T^- depend on the number of carbon atoms in the alkyl group of mononitro or mononitrate alkanes:

(i) if $n = 1$, then $T^- = 1.0$;

(ii) if $n \geq 2$, then $T^- = 0.6$. For $m = 2$, the value of T^- depends on the molecular structure of the energetic compound. The value of T^- equals 0.40 in this case, except if two nitro groups are attached to one carbon, in which case $T^- = 0.0$.

(b) The presence of the –OH group: If the hydroxyl group is present, T^+ is equal to 1.0.

(2) Energetic compounds with –N–NO$_2$, –NH–NO$_2$ and –NHNO$_3$ groups

(a) The presence of –N–NO$_2$ groups: The ratio of the number of –N–NO$_2$ to –CH$_2$ (or –CH$_3$) groups has different effects:

(i) if the ratio $n_{NNO_2}/n_{CH_{2\,or\,3}} \geq 0.5$, then $T^+ = 0.5$;

(ii) if the ratio $n_{NNO_2}/n_{CH_{2\,or\,3}} \leq 0.2$, then $T^- = 0.6$. For other ratios of $n_{NNO_2}/n_{CH_{2\,or\,3}}$, T^+ and T^- equal zero.

(b) The presence of –NH–NO$_2$ and –NHNO$_3$ groups: The value of T^+ is equal to 1.0.

Example 3.4. Consider the following nonaromatic energetic compound:

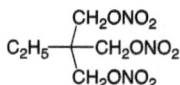

$$
\text{C}_2\text{H}_5\!\!-\!\!\begin{array}{c} \text{CH}_2\text{ONO}_2 \\ \text{CH}_2\text{ONO}_2 \\ \text{CH}_2\text{ONO}_2 \end{array}
$$

The melting point is calculated as follows:

$$
\begin{aligned}
T_m &= 281.7 + 28.97a - 12.08b + 29.75c - 9.966d + 102.92T^+ - 110.11T^- \\
&= 281.7 + 28.97(6) - 12.08(11) + 29.75(3) - 9.966(9) + 102.92(0) - 110.11(0) \\
&= 322.2\,\text{K}.
\end{aligned}
$$

The measured melting point of this compound is 324.15 K [54]. If equation (3.5) is used instead, a value of 293.7 K is obtained, which illustrates that the deviation of (3.5) from the experimental value is larger for such compounds.

3.2.5 Improved method for predicting the melting points of energetic compounds

For various aromatic and nonaromatic energetic compounds containing Ar–NO$_2$, C–NO$_2$, C–ONO$_2$ or N–NO$_2$ groups, an improved method has been introduced which expresses the melting points of these compounds as additive and nonadditive parts. The new correlation has the following form [208]:

$$
T_m = 326.9 + 5.524T_{add} + 101.2T_{nonadd}, \tag{3.7}
$$

where

$$T_{add} = a - 0.5049b + 2.643c - 0.3838d, \tag{3.8}$$

$$T_{nonadd} = T_{PC} - 0.6728T_{NC}. \tag{3.9}$$

Two correcting functions T_{PC} and T_{NC} were chosen based on the deviations of T_{add} from the measured values and are discussed in the following sections.

T_{PC}

(1) $-NH_2$ group: The presence of amino groups can not only enhance the thermal stability of energetic compounds [206], but also can increase their safety by decreasing their sensitivity to external stimuli such as impact [81, 82]. The effect of amino groups on the melting point of a compound can be classified as follows:

(a) The number of amino groups per aromatic ring in polynitro arenes or non-aromatic energetic compounds: If one $-NH_2$ group is present in this type of energetic compound (for example in 2,2′,4,4′,6,6′-hexanitrobiphenyl-3,3′-diamine), the value of T_{PC} is 0.5. One exception to this is o-nitroaniline, because intramolecular hydrogen bonding may cancel the effect of the amino group. The value of T_{PC} equals 2.0 if a larger number of amino groups is present in such compounds, e. g. 1,1-diamino-2,2-dinitroethylene and 1,3,5-tri-amino-2,4,6-trinitrobenzene.

(b) Polynitro heteroarenes: The value T_{PC} equals 1.0 if amino groups are attached to heterocyclic aromatic compounds, e. g. 4-amino-5-nitro-1,2,3-triazole.

(2) The presence of some specific polar groups and molecular fragments: The presence of certain special functional groups such as $-COOH$ can increase the melting points of these compounds because of their ability to form reinforced intermolecular hydrogen bonds. The effects of these functional groups and molecular fragments in different energetic compounds can be classified as follows:

(a) Nitroaromatic: If $-COOH$, $-NH-CO-$ and at least two $-OH$ groups are attached to the aromatic ring, T_{PC} is 0.75. T_{PC} equals 0.75 if $-NO_2$, R_2N- or $-N-C(=O)-$ are present in the *para* position relative to the nitro group. Moreover, the value of T_{PC} is 1.0 if the or $-C=C-Ar$ groups are present.

(b) Nonaromatic energetic compounds: The value of T_{PC} is 1.0 for the existence of some specific polar groups or molecular fragments including $-NH-NO_2$, NH_4^+, more than one $-OH$, one cyclic ether or carbocyclic cage energetic compounds.

T_{NC}

For polynitro arenes and polynitro heteroarenes, the presence of alkyl and alkoxy groups attached to nitroaromatic rings can reduce the planarity of the molecules. The ratio of the number of nitro groups to the number of alkyl or alkoxy groups ($n_{NO_2}/n_{R,OR}$) may be important. For the nitramine polar group, the presence of an alkyl substituent can change its effect, because it can affect the nonplanarity of the nitroaromatic molecules. The attachment of certain heterocycles to a central aromatic ring in poynitro heteroarenes may result in a reduction of the symmetry and planarity of these compounds. Mononitro substitution in aromatic compounds may decrease the melting point compared to polynitro substituted compounds. The effects of various structural parameters on T_{NC} are as follows.

(1) Nitroaromatics: Four different situations can be considered for polynitro arenes and polynitro heteroarenes.
   (a) Alkyl or alkoxy substituted nitroaromatics: For the ratios $n_{NO_2}/n_{R,OR} \leq 1$ and $n_{NO_2}/n_{R,OR} > 1$, the values of T_{NC} equal 1.0 and 0.5, respectively.
   (b) Alkyl nitramine groups attached to aromatic rings: The value of T_{NC} is equal to 0.7 times the number of alkyl nitramine groups.
   (c) Specific structural factors: If ⟨structure⟩, ⟨structure⟩ or ⟨structure⟩ are present, the value of T_{NC} is equal to 1.0.
   (d) Mononitro substituted aromatic compounds: The value of T_{NC} is equal to 1.0.
(2) Nonaromatic energetic compounds: For those nitro and nitrate energetic compounds with general formula $-CH-(NO_2$ or $ONO_2)_n$, the values of T_{NC} are as follows:
   (a) if $n = 1$, then $T_{NC} = 2.0$;
   (b) if $n = 2$ or 3, then $T_{NC} = 1.0$.

Different behavior of some molecular structures

Due to the complex effects of symmetry, planarity and local dipole moments, some specific molecular moieties can have different effects with respect to increasing or decreasing the melting point. The guidelines are as follows:

(1) Polycyclic nitroaromatic compounds: If three aromatic rings are present and if $n_{NO_2}/n_{Ar} > 2.5$, then $T_{PC} = 2.0$. However, if four aromatic rings are present and if $n_{NO_2}/n_{Ar} > 2.5$, then $T_{NC} = 1.0$.
(2) Cyclic nitramines containing methylene units: The ratio of n_{NNO_2} to the number of methylene units (n_{CH_2}) has different effects:
   (a) if the ratio $n_{NNO_2}/n_{CH_2} \geq 1.0$, then $T_{PC} = 0.5$;
   (b) if the ratio $n_{NNO_2}/n_{CH_2} \leq 0.2$, then $T_{NC} = 1.2$.

A summary of the above conditions is given in Tables 3.1 and 3.2.

Table 3.1: Summary of predicted values of T_{PC}.

Energetic compound	Specific groups or molecular moieties	T_{PC}	Comments
Polynitro arene or non-aromatic	−NH$_2$ group (the number of amino groups per ring in polynitro arene)	0.5	One amino group (except *o*-nitroaniline)
		2.0	More than one amino group
Polynitro heteroarene	−NH$_2$ group	1.0	−
Nitroaromatic	−COOH, −NH−CO− and at least two −OH groups −NO$_2$, R$_2$N− or −N−C(=O)− in *para* position with respect to the nitro group [imidazole/pyrazole ring structures] −N, −N or −C=C−Ar	0.75	−
		1.0	
Non-aromatic	−NH−NO$_2$, NH$_4^+$ and more than one −OH as well as one cyclic ether or carbocyclic cage energetic compound	1.0	−
	Cyclic nitramine containing methylene units	0.5	$n_{NNO_2}/n_{CH_2} \geq 1.0$
Polycyclic nitroaromatic	Presence of three aromatic rings	2.0	$n_{NO_2}/n_{Ar} > 2.5$

Table 3.2: Summary of predicted values of T_{NC}.

Energetic compound	Specific groups or molecular moieties	T_{NC}	Comments
Nitroaromatic	Alkyl or alkoxy substituted nitroaromatics	1.0 0.5	$n_{NO_2}/n_{R,OR} \leq 1$ $n_{NO_2}/n_{R,OR} > 1$
	Alkyl nitramine groups attached to aromatic rings	0.7× the number of alkyl nitramine groups	−
	[ring structures: N$^+$−O$^-$ furoxan, furan and N−N ring] Mononitro substituted aromatic compound	1.0	
Non-aromatic	−CH−(NO$_2$ or ONO$_2$)$_n$	2.0 1.0	$n = 1$ $n = 2, 3$
	Cyclic nitramine-containing methylene units	1.2	$n_{NNO_2}/n_{CH_2} \leq 0.2$
Polycyclic nitroaromatic	Presence of four aromatic rings	1.0	$n_{NO_2}/n_{Ar} > 2.5$

Example 3.5. Use equation (3.7) to calculate the melting point of the following compound:

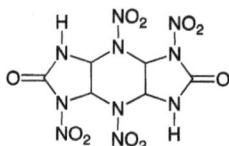

$$T_{add} = a - 0.5049b + 2.643c - 0.3838d$$
$$= 6 - 0.5049(6) + 2.643(10) - 0.3838(10)$$
$$= 25.56$$
$$T_m = 326.9 + 5.524T_{add} + 101.2T_{nonadd}$$
$$= 326.9 + 5.524(25.56) + 101.2(0)$$
$$= 468.1\,K.$$

Since the measured melting point of this compound is 498.15 K [80], the percent deviation of the calculated melting point from the measured melting point (−6.0 %) is much lower than the two group additivity methods of Joback–Reid [175] (1260.8 K, %Dev = 153.1) and Jain–Yalkowsky [194] (772.8 K, %Dev = 55.1).

3.2.6 Organic molecules containing hazardous peroxide groups

The study of various organic compounds containing peroxide groups has shown that it is possible to calculate the melting points of these compounds using core and correcting functions as follows [209]:

$$T_{m,peroxide} = 280.5 + 5.159T_{core} + 38.90T_{correcting}, \tag{3.10}$$
$$T_{core} = a - 0.556b + 2.064d, \tag{3.11}$$
$$T_{correcting} = T^+_{m,peroxide} - 1.345T^-_{m,peroxide}, \tag{3.12}$$

where $T_{m,peroxide}$, T_{core} and $T_{correcting}$ are the melting point of the peroxide compound, and the core and correcting functions, respectively; $T^+_{m,peroxide}$ and $T^-_{m,peroxide}$ are the positive and negative contributions of structural parameters in $T_{correcting}$, respectively. For the presence of several molecular moieties, the values of $T^+_{m,peroxide}$ and $T^-_{m,peroxide}$ can be specified as follows:

(1) $T^+_{m,peroxide}$: The values of $T^+_{m,peroxide}$ are 2.0 and 0.5 for the presence of more than one peroxy acid (without any functional groups) and −OH groups, respectively.

(2) $T^-_{m,peroxide}$: For some organic molecules including those containing −(CO)OO− or −O−C(O)−OO−(CO)−O− groups, the crystal packing efficiency of molecules may

be reduced, which in turn, can decrease the interaction between local dipole moments of neighboring polar groups. If only one $-(CO)OO-$ or $-O-C(O)-OO-(CO)-$ $O-$ group in the form $R_1-(CO)OO-R_1$ or $R_1-O-C(O)-OO-(CO)-O-R_1$ is present (in which R_1 should be the same on both sides of the organic molecule), the value of $T^-_{m,peroxide}$ is 1.0. For $R-C(O)OOH$, the value of $T^-_{m,peroxide}$ is 1.0 if the number of carbon atoms in the R group is less than five.

Example 3.6. Consider the following peroxide:

1,1'-peroxybis(1-hydroperoxycyclohexane)

$$T_{core} = a - 0.556b + 2.064d$$
$$= 12 - 0.556(22) + 2.064(6)$$
$$= 12.2$$
$$T_{m,peroxide} = 280.5 + 5.159T_{core} + 38.90T_{correcting}$$
$$= 280.5 + 5.159(12.2) + 38.90(0)$$
$$= 343.4 \, \text{K}.$$

The measured melting point of this compound is 356 K [142]. The percent deviation of the calculated melting point using the above method is 3.5%, which is much lower than that of the two group additivity methods of Joback–Reid [175] (619 K, %Dev = 74.2) and Jain–Yalkowsky [194] (454 K, %Dev = −27.4).

3.2.7 Organic azides

For organic azides, the following correlation has been introduced [210]:

$$T_m = 264.63 + 10.09a - 3.86b + 18.38c$$
$$- 47.53n_{azide} + 45.76SPG - 65.58IPF, \tag{3.13}$$

where n_{azide} is the number of azide groups; SPG is the contribution of specific polar groups; IPF is the inefficient packing factor. Table 3.3 shows different SPG values containing a list of polar functional groups and molecular fragments. Table 3.4 shows different situations in which the effects of IPF can be considered.

Example 3.7. The use of equation (3.13) for 1-azidonaphthalene ($C_{10}H_7N_3$) gives

$$T_m = 264.63 + 10.09a - 3.86b + 18.38c - 47.53n_{azide} + 45.76SPG - 65.58IPF$$
$$= 264.63 + 10.09(10) - 3.86(7) + 18.38(3) - 47.53(1) + 45.76(0) - 65.58(1.2)$$
$$= 267.4 \, \text{K}.$$

Table 3.3: The values of specific polar groups (*SPG*).

Polar groups	SPG	Exception
–COOH	1.8	Ortho phenoxy molecular moiety (–O–Ar)
–CONH$_2$ or –SO$_2$NH$_2$	2	–
–OH	0.6× the number of –OH groups	Disubstituted benzene containing only –N$_3$ and –OH or –CH$_2$OH
–SO$_3^-$ or	1	–
–NH–CO–NH–, –NH–CO–CH$_3$, cyclic –NH–CO, –CO–NH–N	1.2	–
Cyclic –NH–	0.8	–
Ar–NH$_2$	0.2× the number of –NH$_2$	N–NH$_2$

Table 3.4: The values for the inefficient packing factor (*IPF*).

Compound	Illustration	IPF	Condition
Benzene and naphthalene derivatives	One or two separated benzene rings in form α-azidonaphthalene	1.2	No polar groups are present
	β-azidonaphthalene derivatives, or three separated benzene rings in form two benzene rings in form Ar–Ar	0.6	
R–N$_3$	–	1.8	
Azides containing silicon	–	1.8	No polar groups are present and the silicon atom is directly attached to the azide group
		0.5	No polar groups are present

Since the measured melting point of this compound is 285.16 K [211], the percent deviation of the melting point calculated by this method from that of the measured melting point is 6 %.

3.2.8 General method for the prediction of melting points of energetic compounds including organic peroxides, organic azides, organic nitrates, polynitro arenes, polynitro heteroarenes, acyclic and cyclic nitramines, nitrate esters and nitroaliphatic compounds

Investigations of the effects of different structural features in energetic compounds (including organic peroxides, organic azides, organic nitrates, polynitro arenes, polynitro heteroarenes, acyclic and cyclic nitramines, nitrate esters and nitroaliphatic compounds) on the melting point of a compound has indicated that it is possible to express the melting points of these compounds as a function of additive and nonadditive parts in the following form [212]:

$$T_m = 323.0 + 5.511 T_{add,elem} + 101.2 T_{corr,struc}, \tag{3.14}$$

$$T_{add,elem} = a - 0.5251b + 2.402c, \tag{3.15}$$

$$T_{corr,struc} = T^+_{struc} - 0.6470 T^-_{struc}, \tag{3.16}$$

where $T_{add,elem}$ shows the contribution of the elemental composition as an additive part, whereas the parameter $T_{corr,struc}$ indicates the nonadditive part of the melting point. $T_{corr,struc}$ is a complex function that can be specified on the basis of the molecular structure because some specific polar groups such as –COOH, –NH$_2$ and –OH as well as other molecular moieties may enhance intermolecular interactions (and which are specified later). However, the presence of some specific molecular fragments can decrease the molecular attractions which are present. Large positive and negative deviations of the calculated $T_{add,elem}$ from the experimental data can be adjusted by the parameter $T_{corr,struc}$ in equation (3.14) through the related variables T^+_{struc} and T^-_{struc} in equation (3.16). The values of T^+_{struc} and T^-_{struc} are discussed in the following sections.

3.2.8.1 $T_{corr,struc}$
The contributions of different structural parameters of energetic compounds in $T_{corr,struc}$ can be specified through some specific polar groups and molecular fragments.

T^+_{struc}
(1) Peroxide group: The intermolecular hydrogen bonding effects of –OH, –COOH, –O–OH and COOOH (peroxy acid group) without further functional groups results in the following conditions:
   (a) more than one –O–OH: The value of $T^+_{struc} = 0.5$;
   (b) two or more hydroxyl groups as well as an –O–O– group: $T^+_{struc} = 0.4$;
   (c) the presence of both COOH and COOOH groups: The value of $T^+_{struc} = 0.5$.

(2) Amino group: The melting points of different classes of energetic compounds may be increased by the introduction of amino groups. This can be accounted for as follows:

(a) Nitroarenes and non-aromatic compounds: The value of $T_{struc}^+ = 0.5$ for one $-NH_2$ group (e. g. 2,2′,2″,4,4′,4″,6,6′,6″-nonanitro-1,1′:3′,1″-terphenyl), except for o-nitroaniline. If more than one amino group is present in this category of compound (e. g. 2,2-dinitroethene-1,1-diamine) then the value of $T_{struc}^+ = 2.0$.

(b) Poly- and mono-nitro heteroarenes: The value of $T_{struc}^+ = 1.0$ if amino groups are attached to heterocyclic aromatic compounds (e. g. 7-amino-4,6 dinitrobenzo[c][1,2,5]oxadiazole 1-oxide).

(2) Specific polar groups and molecular fragments: The influence of the molecular fragments and functional groups on different classes of energetic compounds can be classified as follows.

(a) Nitroaromatic energetic compounds: $T_{struc}^+ = 0.75$ if $-COOH$ or $-NH-CO-$ and at least two $-OH$ groups are attached to the aromatic ring. If $-NO_2$, R_2N- and $-N-C(O)-$ groups are attached to the ring in *para* positions with respect to the nitro group, then T_{struc}^+ is also equal to 0.75. The value of $T_{struc}^+ = 1.0$ if the

or $-C=C-Ar$ molecular moieties are present.

(b) Non-aromatic energetic compounds: if $-NH-NO_2$, more than one $-OH$ group, one cyclic ether or carbocyclic cage energetic compound is present, the value of $T_{struc}^+ = 1.0$. Furthermore, $T_{struc}^+ = 0.5$ if NH_4^+ is present in salts of nitro nonaromatic compounds.

T_{struc}^-

For some organic compounds which contain the $-(CO)OO-$ or $-O-C(O)-OO-(CO)-O-$ groups, the $T_{add,elem}$ values are higher than those from experimental data. These groups may reduce the packing efficiency of molecules in crystals because they can decrease the interactions between local dipole moments of neighboring polar groups. The presence of alkyl and alkoxy groups attached to nitroaromatic rings can also reduce the planarity of polynitro arenes and polynitro heteroarene molecules. This nonplanarity not only decreases the packing efficiency of molecules in the crystal lattice, but it can also reduce the interactions between local dipole moments of neighboring nitro groups.

If polar nitramine groups are present, alkyl substituents can reduce the polar effect of these groups because alkyl groups can result in a reduction in the planarity of nitroaromatic molecules. The attachment of heterocycles to the central aromatic ring may result in a reduction of the symmetry in polynitro heteroarenes. Since there are larger attractive forces in polynitro substituted compounds than in mononitro substituted carbocyclic aromatic compounds, the latter have lower melting points. The effects of various structural parameters on T_{struc}^- may be categorized as follows.

(1) The presence of $-(CO)OO-$ or one $-O-O-$ group: If only one $-(CO)OO-$ or $-O-(CO)-OO-(CO)-O-$ is present in the form $R_1-(CO)OO-R_1$ or $R_1-O-(CO)-OO-(CO)-O-R_1$, the value of $T_{struc}^- = 1.0$ (R_1 should be the same on both sides of the organic molecule). The T_{struc}^- value is 1.0 and 0.5 for $R-(CO)OOH$ and $R'-(CO)OH$ (or $R'-OO-R''$), respectively, however, the number of carbon atoms in R and R' should be less than five.

(2) The attachment of $-N_3$ to alkyl or aryl groups: The value of $T_{struc}^- = 1.4$.

(3) Nitroaromatic compounds: The following conditions can be considered for nitro arenes and nitro heteroarenes.

   (a) Alkyl or alkoxy substituted nitroaromatics: For nitroaromatics containing alkyl or alkoxy groups with ratios of $n_{NO_2}/n_{R,OR} \leq 1$ and $n_{NO_2}/n_{R,OR} > 1$, the values of T_{struc}^- are 1.0 and 0.5, respectively.

   (b) Alkyl nitramine groups attached to aromatic rings: $T_{struc}^- = 0.7 \times$ the number of alkyl nitramine groups.

   (c) Specific structural factors in nitroaromatic compounds: If the , or molecular moieties are present, $T_{struc}^- = 1.0$.

   (d) Mononitro substituted aromatic compounds: $T_{struc}^- = 1.0$.

   (e) Nitro heteroarenes: The value of $T_{struc}^- = 0.5$.

(4) Nitro and nitrate nonaromatic energetic compounds: For those nitro and nitrate energetic compounds with general formula $-CH-(NO_2$ or $ONO_2)_n$:

   (a) if $n = 1$, then $T_{struc}^- = 2.0$;

   (b) if $n = 2$ or 3, then $T_{struc}^- = 1.0$.

Polycyclic nitroaromatic compounds as well as cyclic nitramines with methylene units

In polycyclic nitroaromatic compounds as well as cyclic nitramines which contain methylene units, there are complex effects related to symmetry, planarity, and local dipole moments. Thus, the following conditions should be applied:

(1) Polycyclic nitroaromatic compounds: If three or four aromatic rings are present, the ratio of n_{NO_2} to n_{Ar} can estimate the contribution of T_{struc}^+ and T_{struc}^-. If $n_{NO_2}/n_{Ar} > 2.5$, the value of T_{struc}^+ is 2.0 if three aromatic rings are present, whereas if four aromatic rings are present, $T_{struc}^- = 1.0$ if $n_{NO_2}/n_{Ar} > 2.5$.

(2) Cyclic nitramines containing methylene units: Two situations are encountered based on the ratio of n_{NNO_2} to the number of methylene units (n_{CH_2}):

   (a) if the ratio $n_{NNO_2}/n_{CH_2} \geq 1.0$, $T_{struc}^+ = 0.5$;

   (b) if the ratio $n_{NNO_2}/n_{CH_2} \leq 0.2$, $T_{struc}^- = 1.2$.

Tables 3.5 and 3.6 summarize the predicted values of T_{struc}^+ and T_{struc}^- which are obtained by applying the above conditions.

Table 3.5: Summary of predicted values of T_{struc}^+.

Energetic compound	Specific groups or molecular moieties	T_{struc}^+	Condition	Example
Peroxide	O–OH group	0.5	More than one hydroperoxy group	2,5-dihydroperoxy-2,5-dimethylhexane
	OH group	0.4	Two or more hydroxy groups	peroxydimethanol
	COOH and COOOH groups	0.5	The presence of two similar groups	2-carboperoxybenzoic acid
Nitro arenes or non-aromatics containing amino groups	–NH$_2$ group	2.0	More than one amino group	2,4,6-trinitrobenzene-1,3-diamine 2,2-dinitroethene-1,1-diamine
	–NH$_2$ group	0.5	One amino group (except o-nitroaniline)	p-Nitroaniline
Poly and mononitro heteroarene	–NH$_2$ group	1.0	–	5-nitro-2H-1,2,3-triazol-4-amine 7-amino-4,6-dinitrobenzo[c][1,2,5]oxadiazole 1-oxide
Nitro aromatic	–COOH, –NH–CO, at least two –OH groups, –NO$_2$, R$_2$N– or –N–C(O)– in para position relative to the nitro group	0.75	–	2,4,6-Trinitroresorcinol Benzenamine, N,N-dimethyl-4-nitro-
	and C=C–Ar	1.0	–	N-(2,4,6-trinitrophenyl)-1H-1,2,4-triazol-3-amine

Table 3.5 (continued)

Energetic compound	Specific groups or molecular moieties	T^+_{struc}	Condition	Example
Nitro non-aromatic	Cyclic nitramines containing methylene units	0.5	$n_{NNO_2}/n_{CH_2} \geq 1.0$	2,4,6-Trinitroresorcinol
	−NH−NO$_2$ and more than one −OH, as well as one cyclic ether group	1.0	−	N,N'-(ethane-1,2-diyl)dinitramide
	NH$_4^+$	0.5	−	NH4+ N-nitronitramide, ammonium salt
Polycyclic nitro-aromatics	The presence of three aromatic rings	2.0	$n_{NO_2}/n_{Aromatic\ ring} >$ 2.5	3,5-dinitro-N2,N6-bis(2,4,6-trinitrophenyl)pyridine-2,6-diamine

Example 3.8. The use of this method for the following compound

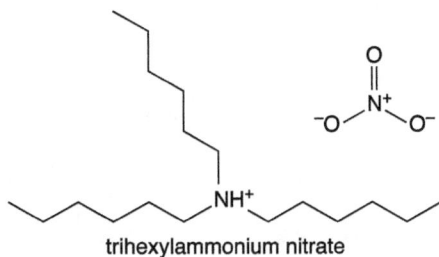

trihexylammonium nitrate

Table 3.6: Summary of predicted values of T_{struc}^-.

Energetic compound	Specific groups or molecular moieties	T_{struc}^-	Condition	Example
Peroxide	O–OH group	0.5	One hydroperoxy group	2-hydroperoxy-2-methylpropane
	O–O group	0.5	One group without any other group present	(2-(*tert*-butylperoxy)propan-2-yl)benzene
	(CO)OH, (CO)OO group	1.0	Only one group without other groups present	*tert*-butyl 2,2-dimethylpropaneperoxoate
Azide	N_3	1.4	The presence of N_3 group except ionic azides	1-azido-1,2,2-trimethylcyclopropane
Nitro non-aromatic	$-CH-(NO_2$ or $ONO_2)_n$	2.0	$n = 1$	Ethyl Nitrate
		1.0	$n = 2, 3$	4-nitrobutan-2-yl nitrate
	Cyclic nitramines containing methylene units	1.2	$n_{NNO_2}/n_{CH_2} \leq 0.2$	Piperidine, 1-nitro
Nitroaromatic	Alkyl or alkoxy substituted nitroaromatics	1.0	$n_{NO_2}/n_{R,OR} \leq 1$	Benzene, 2,4-dimethyl-1-nitro
		0.5	$n_{NO_2}/n_{R,OR} > 1.0$	Benzene, 2-methyl-1,3,5-trinitro-

Table 3.6 (continued)

Energetic compound	Specific groups or molecular moieties	T^-_{struc}	Condition	Example
	Alkyl nitramine groups attached to aromatic rings	$0.7 \times A$	A is the number of alkyl nitramine groups	N-Methyl-N,2,4,6-tetranitroaniline
	Mononitro substituted aromatic compounds	1.0	–	Benzene, 1-methyl-2-nitro-
	The presence of one NO_2 group in two ring-fused heteroarenes	0.5	–	5-nitroquinoline
		1.0		7-amino-4,6-dinitrobenzo[c][1,2,5]oxadiazole 1-oxide
Polycyclic nitro-aromatic	$n_{NO_2}/$ $n_{Aromatic\ ring} > 2.5$	1.0	The presence of four aromatic rings	2,4,6-trinitro-N1,N3,N5-tris(2,4,6-trinitrophenyl)benzene-1,3,5-triamine

predicts the melting point as

$$T_{add,elem} = a - 0.5251b + 2.402c$$

$$= 18 - 0.5251(39) + 2.402(2)$$

$$= 2.325,$$

$$T_m = 323.0 + 5.511T_{add,elem} + 101.2T_{corr,struc}$$

$$= 323.0 + 5.511(2.325) + 101.2(0)$$

$$= 335.8 \text{ K}.$$

The predicted melting point of this compound is close to the experimentally determined value of 345 K [142].

3.2.9 Cyclic saturated and unsaturated hydrocarbons

Cyclic hydrocarbons or alicyclic hydrocarbons are a kind of closed chain hydrocarbons where they and their derivatives are cyclic hydrocarbons of alkane, alkene, and/or alkyne, which contain at least one ring. Polycyclic compounds are important cyclic hydrocarbons with more than one ring of carbon atoms whose rings share two or more same carbon atoms. The heats of combustion of cyclic hydrocarbons are much larger than we might expect by analogy with noncyclic compounds because of the existence of strain energy in cyclic hydrocarbons [6]. Thus, polycyclic saturated hydrocarbons can be used as liquid fuels, which are used in liquid bipropellants [1–3, 7]. For example, RJ-4 is a mixture of *endo*- and *exo*-tetrahydrodimethyldicyclopentadiene that is used as a synthetic jet fuel [7].

A general correlation has been developed for the prediction of the melting point of a wide range of cyclic saturated and unsaturated hydrocarbons with and without substituents including cyclic alkane, alkene, and/or alkyne, cage molecules, bridged cyclic and multicyclic hydrocarbon structures [213]. It depends on three variables as follows [213]:

$$T_m = 103.1 + 10.24a + 89.16T^+_{m,cyc\,hyd} - 66.44T^-_{m,cyc\,hyd}, \tag{3.17}$$

where $T^+_{m,cyc\,hyd}$ and $T^-_{m,cyc\,hyd}$ are two correction terms, which correspond to increasing and decreasing values of the melting point corresponding to specific structural parameters given in Tables 3.7 and 3.8.

The present method can provide a suitable correlation for the quick reliable finding of novel cyclic hydrocarbon fuels with desirable melting points.

Table 3.7: Summary of the values of $T^+_{m,cyc\ hyd}$.

Type of cyclic hydrocarbon	Condition	$T^+_{m,cyc\ hyd}$	Example
Saturated (without substituents)	$5 < n_{C,ring} * < 19$ for monocyclic ring	1.0	
	Bicyclic ring	1.5 (for fused rings where each ring containing less than a five-membered ring)	
		2.0 (for bridged rings where each ring containing more than four-membered ring)	
	Tricyclic ring	1.5 (*Endo*)	
		3.0 (symetric consecutive ring)	
	Cage hydrocarbon with more than bicyclic ring	1.7	
		3.5 (more than a four-membered ring)	
	Polycyclic containing more than three rings where each ring directly attached to another ring	1.0	
Saturated (with substituents)	$n_{C,ring} = 6$ with *trans*-dialkyl similar substituents or *cis,cis*-trialkyl with more than one similar substituents	1.0	
	The bridged bicyclic ring where each ring containing more than four-membered ring	1.7	
		1.0 (one substituent)	
	Cage hydrocarbon containing more than two rings	1.7	
		2.2 (one substituent and more than a four-membered ring)	
	Each carbon of ring containing one or two methyl substituents	0.8	

Table 3.7 (continued)

Type of cyclic hydrocarbon	Condition	$T^+_{m,cyc\ hyd}$	Example
Unsaturated (without substituents)	$n_{C,ring}$ = even number and number of double bonds = $\frac{n_{C,ring}}{2}$ or including only *trans* isomers with one double bond or allene	0.8	
	The bridged bicyclic ring where each ring containing more than four-membered ring	2.0 (one double bond)	
		1.0	
	Endo-tricyclic ring	1.5	
	Directly attachment of two rings with more than six carbon atoms in each ring	1.0	
	$n_{C,ring}$ > 11 in which all double bonds are in *trans* form	1.0	
	The existence of $-C \equiv C-$ in ring	1.2	
Unsaturated (containing substituents)	Bridgedbicyclic ring with an only methyl substituent	2.0	
	Attachment of substituents to ring through a double bond	1.0	

Example 3.9. 1,3,3-Trimethyl-bicyclo[2.2.1]heptane has the following molecular structure:

Table 3.8: Summary of the values of $T_{m,cyc\ hyd}^{-}$.

Type of cyclic hydrocarbon	Condition	$T_{m,cyc\ hyd}^{-}$	Examples
Saturated (without substituents)	$n_{C,ring}^{*} > 26$	1.0	
Saturated (containing substituents)	One substituent with a nonlinear chain with more than six carbons or the presence of a molecular fragment ⁓⁓⁓ where R contains more than five carbon atoms	1.2	Cyclic alkane / Cyclic alkane —R
	The existence of molecular moiety R—⟨ Cyclic alkane / Cyclic alkane	1.7	

$^{*}n_{C,ring}$ = The number of carbon atom in the ring.

The use of equation (3.17) gives:

$$T_m = 103.1 + 10.24a + 89.16T_{m,cyc\ hyd}^{+} - 66.44T_{m,cyc\ hyd}^{-}$$
$$= 103.1 + 10.24 \times 10 + 89.16 \times 1.7 - 66.44 \times 0$$
$$= 357.1\,K$$

The experimental value of the melting point of this compound is 331.0 K [142]. Meanwhile, the use of the method by Alantary and Yalkowsky [202] – as one of the best available predictive methods – gives 211.4 K.

3.3 Melting points of ionic liquids

Ionic liquids are low melting salts, which melt at or near room temperature. They have wide applications in the chemistry and chemical industries [214, 215]. Melting points of ionic liquids are essential for designs and applications but their melting-point data are scarce [216]. They show electrostatic attraction because larger sizes of cations and anions have a larger distance between two ions, which reduces their electrostatic attraction [217]. They depend on the arrangement of ions and the strength of the pairwise ion interactions in the crystal. The structure and interactions between the ions can provide ionic liquids with low melting points. Different approaches have been developed for the prediction of melting points of selected classes of ionic liquids – quantitative structure-property relationships (QSPR) and group additivity methods are two common methods [218, 219]. Molecular dynamics (MD) and density functional theory

(DFT) can predict melting points by the simulation of solid-to-liquid phase transitions but they are difficult as compared to other estimation methods [220, 221]. The high computational demands and the need for starting geometry are the main limitations of MD and DFT methods. Low et al. [218] investigated the effect of descriptor choice in machine learning models for ionic liquid melting point prediction from an experimental dataset of 2212 ionic liquid melting points consisting of diverse ion types. They introduced the best model, based on ECFP4 and molecular orbital energies, for the prediction of ionic liquid melting points with an average mean absolute error of 29 K.

Chen and Bryantsev [220] used complex dispersion-corrected DFT method involving (semi)local (PBE-D3) and hybrid exchange-correlation (HSE06-D3) functionals to predict the lattice enthalpy, entropy, and free energy of 11 ILs containing imidazolium/pyrrolidinium cations and halide/polyatomic fluoro-containing anions.

Venkatraman et al. [219] used various machine learning methods to predict the melting points of structurally diverse 2212 ILs based on a combination of 1369 cations and 141 anions. They found that tree-based ensemble methods (Cubist, random forest and gradient boosted regression) can demonstrate slightly better performance over support vector machines and k-nearest neighbor approaches.

Mehrkesh and Karunanithi [221] used the quantum chemistry descriptors of the cation and anion radii, the density of ionic liquids, symmetrical value, the molecular weight of cation, and the dielectric energy of anion for the prediction of melting points of ILs based on the 37 experimental melting point data.

3.3.1 Group additivity approach

The group additivity methods can be used only for ionic liquids containing the specified cations and anions. Thus, they cannot be applied for many new ionic liquids where the contributions of cation and anion groups have not been considered. Lazzús [222] used experimental data of melting points of 40 ionic liquids to obtain the contributions of the cation-anion groups in a correlation set. Another data set including 23 ionic liquids was used to test the reliability of the method. The final equation for this model is [222]:

$$T_m = 98.599 + \sum_i n_{c,i} T_{c,i} + \sum_j n_{a,j} T_{a,j}, \tag{3.18}$$

where $n_{c,i}$ and $n_{a,j}$ are the occurrences of the groups i and j in the cation and the anion of the desired ionic liquid, respectively; $T_{c,i}$ and $T_{a,j}$ are the contribution of the cation and anion groups, respectively. Cation groups contain imidazolium, pyridinium, pyrrolidinium, phosphonium, and ammonium. Anion groups contain halides, pseudohalides, sulfonates, tosylates, imides, borates, phosphates, carboxylates, and metal complexes. The contribution values of the mentioned cation and anion groups are given in Table 3.9.

Table 3.9: The contribution of cation and anion groups for calculation of melting points of ionic liquids.

Cations

Name	Group	$T_{c,i}$	Group	$T_{c,i}$
Imidazolium		39.698	Z=–H	38.623
Pyridinium		82.227	Z=–CH$_3$	68.819
Pyrrolidinium		12.868	Z=–CH$_2$–	1.344
Ammonium		13.890	Z=–CH<	–79.375
Phosphonium		–448.738	Z=–N	21.626

Anions

Group	$T_{a,j}$	Group	$T_{a,j}$
=CH–	21.765	–Cl	10.923
>C<	9.910	–Br	10.213
–COO	13.484	–P [>P<]	33.726
–HCOO	13.531	–B [>B<]	–20.084
–O– [–O]	–9.850	–I	–3.753
–N– [>N–]	–5.493	>As<	–28.174
–NO$_3$	–4.482	–CB$_{11}$H$_6$	–5.603
–SO$_2$–	8.757	–CB$_{11}$H$_{12}$	66.553
–CF$_3$	–41.448	=CH– (ring)	8.067
–CF$_2$–	–1.811	=C< (ring)	3.132
–F	–4.930		

Example 3.10. Consider the following ionic liquid:

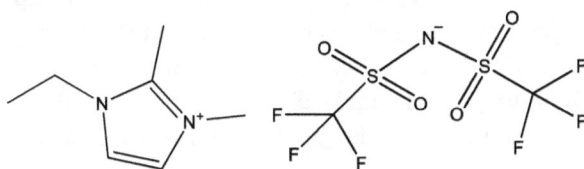

The melting point is calculated as follows:

$$T_m = 98.599 + \sum n_{c,i} T_{c,i} + \sum n_{a,j} T_{a,j}$$
$$= 98.599 + \text{Imidazolium} + 3(T_{c,\text{-CH}_3}) + T_{c,\text{-CH}_2\text{-}} + T_{a,\text{-N-}} + 2(T_{a,\text{-SO}_2}) + 2(T_{a,\text{-CF}_3})$$

$$= 98.599 + 39.698 + 3(68.819) + 1.344 - 5.493_{+}2(8.757) + 2(-41.448)$$
$$= 275.22 \, \text{K}.$$

The measured melting point of this compound is 248.15 K [223].

3.3.2 QSPR approaches based on complex descriptors

QSPR approaches usually need complex descriptors, expert users, and computer codes to evaluate the melting point of the desired compound. They correlate the requested chemical or physical properties with molecular descriptors, which are derived from the molecular structures of chemical compounds quantitatively [224]. They are based on different kinds of statistical tools with some restrictions, e. g. application to similar compounds and the correct selection of molecular descriptors. They can predict the melting points of different classes of organic compounds. Wang et al. [225] used multiple linear regression and an artificial neural network for the estimation of the melting points of carbocyclic nitroaromatic compounds with six complex descriptors. Liu and Holder [226] have developed three QSPR models for organosilicon compounds containing a variety of silicon-containing organic compounds, silanes, and siloxanes, respectively. Liang et al. [227] have correlated the melting points of fatty acids with five complex descriptors, e. g. the average valence connectivity index chi-5. Yan et al. [228] developed a QSPR model for the prediction of melting points of eight kinds of ionic liquid compounds including imidazolium, benzimidazolium, pyridinium, pyrrolidinium, ammonium, sulfonium, triazolium, and guanidinium based on the general topological index. Watkins et al. [229] used 49 complex descriptors to estimate the melting points of a wide range of persistent organic pollutants including chloro- and bromo-analogs of dibenzo-p-dioxins, dibenzofurans, biphenyls, naphthalenes, diphenyl ethers, and benzenes by GA-partial least squares (GA–PLS) modeling and the random forest method. Morrill and Byrd [230] developed some QSPR models to predict melting temperatures of energetic materials based on descriptors calculated using the AM1 semiempirical quantum mechanical method. Farahani et al. [231] developed a QSPR model using 12 complex descriptors such as descriptor X1A, which stands for average connectivity index of order 1, and genetic function approximation for the prediction of melting points of diverse ionic liquid compounds including sulfonium, ammonium, pyridinium, 1,3-dialkyl imidazolium, tri-alkyl imidazolium, phosphonium, pyrrolidinium, double imidazolium, 1-alkyl imidazolium, piperidinium, pyrroline, oxazolidinium, amino acids, guanidinium, morpholinium, isoquinolinium, and tetra-alkyl imidazolium.

3.3.3 Simple approach based on the structure of cations and anions

A simple correlation has been developed to predict melting points of important classes of ILs including imidazolium-, pyridinium-, pyrrolidinium-, ammonium-, phosphonium-, and piperidinium-based ILs and different type of anions. The experimental data of 195 different types of ILs were used to derive the new correlation as [232]:

$$T_m = 382.1 - 13.35a_{cat} + 4.796b_{cat} - 22.07c_{cat} - 11.24b_{ani} - 25.29c_{ani}$$
$$+ 70.69j_{ani} + 42.37f_{ani} - 205.4g_{ani} + 116.7T^+_{m,IL} - 121.8T^-_{m,IL} \qquad (3.19)$$

where a_{cat}, b_{cat}, and c_{cat} are the number of carbon, hydrogen, and nitrogen atoms in cation, respectively; b_{ani}, c_{ani}, j_{ani}, f_{ani}, and g_{ani} are the number of hydrogen, nitrogen, bromine, chlorine and aluminum atoms in anion, respectively; $T^+_{m,IL}$ and $T^-_{m,IL}$ are positive and negative adjusting functions, respectively. Table 3.10 shows the values of $T^+_{m,IL}$ and $T^-_{m,IL}$ for the presence of specific cations and anions in the desired IL.

Example 3.11. Consider the following ionic liquid:

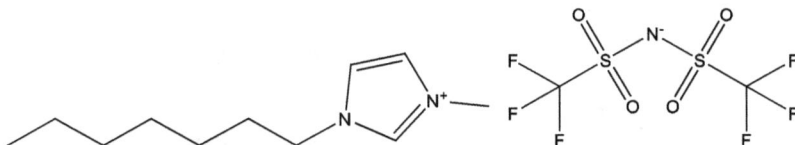

The use of equation (3.19) gives:

$$T_m = 382.1 - 13.35a_{cat} + 4.796b_{cat} - 22.07c_{cat} - 11.24b_{ani} - 25.29c_{ani}$$
$$+ 70.69j_{ani} + 42.37f_{ani} - 205.4g_{ani} + 116.7T^+_{m,IL} - 121.8T^-_{m,IL}$$
$$= 382.1 - 13.35 \times 11 + 4.796 \times 21 - 22.07 \times 2 - 11.24 \times 0 - 25.29 \times 1$$
$$+ 70.69 \times 0 + 42.37 \times 0 - 205.4 \times 0 + 116.7 \times 0 - 121.8 \times 0 = 266.5K$$

The measured melting point of this ionic liquid is 219.6 K [233].
   The use of equation (3.18) for this ionic liquid provides:

$$T_m = 98.599 + \sum n_{c,i}T_{c,i} + \sum n_{a,j}T_{a,j}$$
$$= 98.599 + \text{Imidazolium} + 2(T_{c,\text{-CH}_3}) + T_{c,\text{-H}} + 6(T_{c,\text{-CH}_2\text{-}}) + T_{a,-N-} + 2(T_{a,\text{-SO}_2})$$
$$+ 2(T_{a,\text{-CF}_3})$$
$$= 98.599 + 39.698 + 2(68.819) + 38.623 + 6(1.344) - 5.493 + 2(8.757) + 2(-41.448)$$
$$= 251.89 \text{ K}$$

3.4 Summary

Several empirical methods have been introduced for predicting the melting points of organic compounds containing energetic functional groups. For polynitro arenes and polynitro heteroarenes, the reliability of equation (3.4) may be better than that of (3.3).

Table 3.10: The values of $T^+_{m,IL}$ and $T^-_{m,IL}$ for the presence of specific cations and anions in the desired IL.

Cation	Anion	$T^+_{m,IL}$	$T^-_{m,IL}$	Condition
imidazolium $H_3C\text{-}(CH_2)_n$ (methyl) or $H_3C\text{-}(CH_2)_n$ (ethyl)	BF_4^-	0.5	0.0	$n = 0$
		0.0	0.80	$2 < n < 10$
		0.5	0.0	$n > 11$
	PF_6^-	0.0	0.5	$5 < n < 10$
		0.5	0.0	$n > 11$
	$\text{—}C_6H_4\text{—}SO_3^-$	11.0	0.0	$n < 4$
	Cl^-	0.0	0.5	$2 < n < 10$
	trifluoroacetate	0.0	0.3	$1 < n < 10$
		0.5	0.0	$n > 11$
	perfluoro-borate	0.0	0.5	$1 \leq n < 10$
	$(F_3C\text{-}SO_2)_2N^-$	0.5	0.0	$n > 11$
dimethylimidazolium $(CH_2)_n$	Cl^-	10.8	0.0	$0 < n < 4$
	trifluoroacetate	0.7	0.0	$n < 3$
methyl-pyridinium $(CH_2)_n$	Br^-	1.1	0.0	$3 < n < 7$
	$(F_3C\text{-}SO_2)_2N^-$	0.7	0.0	
pyridinium $(CH_2)_n$	PF_6^-	1.0	0.0	$11 < n < 18$
pyrrolidinium $(CH_2)_n$	$(F_3C\text{-}SO_2)_2N^-$	0.5	0.0	$n < 3$
	phthalate or phosphate	0.6	0.0	$n < 4$
	perfluoro-borate	0.0	0.5	$n < 4$
	$\text{—}SO_3^-\,C_4F_9$ or sulfonate	11.0	0.0	$n = 0$
		0.4	0.0	$1 \leq n < 8$

Table 3.10 (continued)

Cation	Anion	$T^{+}_{m,\mathrm{IL}}$	$T^{-}_{m,\mathrm{IL}}$	Condition
	(aryl)–SO₃⁻	11.0	0.0	$n < 6$
	NC–N⁻–CN (dicyanamide)	11.0	0.0	$n = 1$
	methylsulfate	11.3	0.0	
ammonium, –(CH₂)ₘ	F₃C–SO₂–N⁻–SO₂–CF₃	0.5 / 0.0	0.0 / 0.3	$m < 3$ / $3 < m$
quaternary ammonium		0.6	0.0	$n < 5$
		0.2	0.0	$n \geq 5$
	(aryl)–SO₃⁻	11.0	0.0	
ammonium, –(CH₂)ₙ / –(CH₂)ₘ	BF₄⁻	0.7	0.0	m or $n < 4$
	F₃C–SO₂–N⁻–SO₂–CF₃	0.0	0.5	
piperidinium, –(CH₂)ₘ	BF₄⁻	0.5	0.0	$n < 9$
phosphonium, –(CH₂)ₙ	F₃C–SO₂–N⁻–SO₂–CF₃	0.4	0.0	$n < 3$
phosphonium, –(CH₂)ₙ	Br⁻	0.0	0.5	n or $m < 7$

For nonaromatic energetic compounds, the use of equation (3.6) is recommended rather than equation (3.5). Equations (3.7) to (3.9) can be applied to various aromatic and non-aromatic energetic compounds containing $Ar-NO_2$, $C-NO_2$, $C-ONO_2$ or $N-NO_2$ groups. Equations (3.10) to (3.12) and (3.13) can be used to predict the melting points of those energetic compounds which only contain peroxide and azide groups, respectively. Equations (3.14) to (3.16) provide a general approach for the prediction of melting points of various energetic compounds, including those of organic peroxides, organic azides, organic nitrates, polynitro arenes, polynitro heteroarenes, acyclic and cyclic nitramines, nitrate esters and nitroaliphatic compounds, aromatic ring, *ISSP* equals 1.0. Equation (3.17) provides a reliable correlation for prediction of the melting points of cyclic hydrocarbons, which can be used as jet fuels. Several have also been introduced for prediction of melting points of ILs. Equation (3.18) is based on group additivity approach, which can be applied for those ILs containing cations with imidazolium, pyridinium, pyrro-lidinium, phosphonium, and ammonium groups as well as anions with halides, pseudohalides, sul-fonates, tosylates, imides, borates, phosphates, carboxylates, and metal complexes. Equation (3.19) can be used for a wide range of ILs including imidazolium-, pyridinium-, pyrrolidinium-, ammonium-, phosphonium-, and piperidinium-based ILs and different type of anions.

4 Enthalpy and entropy of fusion

Thermal analysis of energetic compounds shows that they decompose at specific temperatures. Their exothermic decomposition can inhibit the explosive charge from being able to dissipate the applied heat [234]. The decomposition reaction begins usually above or during the melting process, so that energetic materials with higher melting points show high thermal stability. The enthalpy – or heat of fusion ($\Delta_{fus}H$) – is the enthalpy change which occurs in the transition from the most stable form of the solid to the liquid state of high energy compounds. It is related to the entropy of fusion ($\Delta_{fus}S$) and the melting point or fusion temperature, i. e. $\Delta_{fus}H = T_m\Delta_{fus}S$. To measure the $\Delta_{fus}H$ of explosive materials, differential scanning calorimetry (DSC) can be used [235]. In this chapter, different approaches for the prediction of $\Delta_{fus}H$ and $\Delta_{fus}S$ will be reviewed.

4.1 Different approaches for the prediction of the enthalpy of fusion

Zeman et al. [236–239] have introduced some relationships between $\Delta_{fus}H$ and the impact sensitivity – as well as with the electric spark sensitivity – of nitramines and polynitro compounds. Therefore, the prediction of $\Delta_{fus}H$ provides a better insight into the intermolecular interactions and sensitivity of energetic molecules which have not yet been synthesized.

QSPR [240], quantum mechanics [241], group additivity methods [242, 243], artificial neural networks [244] and simple correlations based on molecular structures [245–251] are all suitable methods for predicting $\Delta_{fus}H$. QSPR methods need special computer codes and the databank set should cover a large number of compounds with different molecular structures in order to obtain suitable results for compounds with similar molecular structures as those included in the test library. Furthermore, they require complex molecular descriptors. However, such methods have often been used to predict the thermodynamic properties of particular families of compounds [252]. The group additivity methods have also been developed to predict the values of $\Delta_{fus}H$ for different types of organic compounds [187, 253–256]. However, it has been shown that they may give $\Delta_{fus}H$ values which show large deviation from the expected

https://doi.org/10.1515/9783110740158-004

values for some organic energetic compounds [188, 240]. Quantum mechanical methods were used to study the phase change properties of some energetic compounds [241, 257, 258], however, they require high-speed computers and specific computer codes. Some simple methods have also been developed for the prediction of $\Delta_{fus}H$ values [245–249] based on the molecular structures of energetic compounds. The group additivity method of Yalkowsky and coworkers [199–202], as well as simple methods based on molecular structures, will be demonstrated in the following sections.

4.1.1 Group additivity method for prediction of the enthalpy of fusion

The UPPER approach of Yalkowsky et al. [199–202] treats enthalpic interactions in additive constitutive properties in the general form:

$$\Delta_{fus}H = \sum n_i H_{fus,i}, \tag{4.1}$$

where $\Delta_{fus}H$ is the enthalpy of fusion of the whole molecule, n_i is the number of fragment i in the molecule, and $H_{fus,i}$ is the contribution of molecular fragment or group i to the enthalpy of fusion. Table 4.1 shows the values of $H_{fus,i}$ for different molecular fragments.

Example 4.1. Consider 2,4-dinitrobenzoic acid with the following molecular structure:

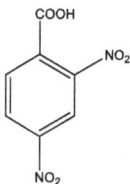

The enthalpy of fusion of this compound using equation (4.1) and Table 4.1 is given as follows:

$$\Delta_{fus}H = \sum n_i H_{fus,i} = 3C_{ar} + 3CH_{ar} + 2YNO_2 + YCOOH$$
$$= 3 \times 1.777 + 3 \times 1.235 + 2 \times 4.584 + 11.785 = 29.99 \text{ kJ mol}^{-1}$$

The reported enthalpy of fusion for this compound is 24.6 kJ mol^{-1} [142].

As seen in Table 4.1, the groups that participate in hydrogen bonds have higher $H_{fus,i}$ values than those containing only dipole–dipole bonds because the latter only participates in dispersion forces.

Table 4.1: The contribution of i group or fragment in calculation of the enthalpy of fusion.

Group	Contribution	Group	Contribution	Group	Contribution
XCH_3	0.701	CH_{fus}	1.695	XNH_2	6.884
YCH_3	1.221	C_{fus}	1.332	XNH	1.181
ZCH_3	0.331	$CH_2 RING$	1.054	YNH_2	5.681
XCH_2	1.408	CH_{RING}	1.046	YNH	3.799
XCH_2*	−2.524	C_{RING}	0.757	YN	2.013
YCH_2	1.644	$=CH_{RING}$	0.883	YNO_2	4.584
$YYCH_2$	1.875	$=C_{RING}$	1.362	$YNHCO$	8.167
$YZCH_2$	−0.976	XF	−0.87	XCN	5.174
XCH	1.177	YF	0.409	YCN	6.558
YCH	−0.916	XCl	1.889	$XCOO$	9.488
XC	1.177	YCl	1.581	$YCOO$	6.208
YC	−1.076	XBr	4.674	$XCOOH$	14.287
$CH_2=$	0.454	$2\&6$	−2.954	$YCOOH$	11.785
$YCH=$	1.691	YBr	2.911	$YCHO$	5.470
$YYCH=$	1.689	XI	4.034	XCO	8.037
$YC=$	2.250	YI	4.334	YCO	3.332
$YYC=$	0.655	XO	2.921	$YOCO$	7.568
$CH\equiv$	2.357	YO	3.162	$YOCOO$	5.335
$ZC\equiv$	3.853	YYO	−6.918	$YCONH_2$	12.814
$YZC\equiv$	−1.732	$Ar-O$	−0.922	$YCONH$	9.083
$C_{allenic}$	2.033	XOH	4.953	$YNHCOO$	6.929
C_{ar}	1.777	YOH	6.699	$YNHCONH_2$	14.865
CH_{ar}	1.235	YSH	2.635	$YNHCON$	16.721
C_{BIP}	2.602	YS	5.313	$YCONH_2$	13.418
C_{BR1}	1.329	YSO_2NH_2	10.642	Ortho	−0.282
C_{BR2}	−0.564	YSO_2N-X	6.739	IHB	−3.495

X = A group that is bonded to only sp^3 atoms.
Y = A group that is singly bonded to one sp^2 atom.
YY = A group bonded to 2 sp^2 atoms.
Z = A group that is bonded to a sp atom.
RING = A group within an aliphatic ring.
fus = An aliphatic bridge-head group.
ar = A group within an aromatic ring.
BR2 = An aromatic carbon contained in 2 rings.
BR3 = An aromatic carbons contained in 3 rings.
BIP = The central carbons of biphenyl rings.
Ortho = The number of ortho substitutions.
2&6 = The number of halogen substitutions at the 2 and 6 positions of a biphenyl ring systems.

4.1.2 Nitroaromatic carbocyclic energetic compounds

The study of the $\Delta_{fus}H$ values for various nitroaromatic carbocyclic compounds with general formula $C_aH_bN_c(O \text{ or } S)_{d''}$ has shown a new approach can be used to derive a useful equation for predicting $\Delta_{fus}H$ as follows [247]:

$$\Delta_{fus}H = 1.197 + 1.681a + 6.793c - 2.143d'' + 8.526C_{SPG}, \tag{4.2}$$

where $\Delta_{fus}H$ is in kJ/mol and C_{SPG} is the contribution of specific polar groups attached to an aromatic ring. As is seen in equation (4.2), the C_{SPG} coefficient has a positive sign which can result in an increase in the value of $\Delta_{fus}H$. The value of C_{SPG} is determined as follows.

(1) Hydroxyl groups: C_{SPG} corresponds to the number of hydroxyl groups attached to an aromatic ring, e. g. $C_{SPG} = 2.0$ for 1,3-dihydroxy-2,4,6-trinitrobenzene.

(2) Amino group: The value of C_{SPG} equals 2.0 for nitroaromatic carbocyclic compounds which have more than two amino ($-NH_2$) groups attached to the aromatic ring, e. g. $C_{SPG} = 2.0$ for 1,3,5-triamino-2,4,6-trinitrobenzene (TATB).

(3) Other polar groups: The value of C_{SPG} is 1.0 if at least one $-C(=O)-O-C(=O)-$ or $-S(O)_2-$ (sulfone) is attachd to an aromatic ring, whereas C_{SPG} is 2.0 if at least one $-COOH$ functional group is attached to an aromatic ring. For example, $C_{SPG} = 1.0$ for 4-nitrophthalic anhydride and $C_{SPG} = 2.0$ for 2-nitrobenzoic acid.

(4) Disubstituted nitroaromatic compounds: The value of C_{SPG} is 1.0 for disubstituted nitroaromatic compounds that contain only two nitro groups attached to aromatic rings, e. g. 1,8-dinitronaphthalene.

Example 4.2. The use of equation (4.2) for 2,2′,4,4′,6,6′-hexanitrobiphenyl with the following structure gives

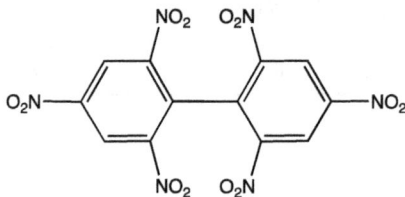

$$\Delta_{fus}H = 1.197 + 1.681a + 6.793c - 2.143d'' + 8.526C_{SPG}$$
$$= 1.197 + 1.681(12) + 6.793(6) - 2.143(12) + 8.526(0)$$
$$= 36.41 \, kJ/mol$$

The predicted value is close to the experimental value of 37.44 kJ/mol [237].

4.1.3 Nitroaromatic energetic compounds

Since equation (4.2) may show relatively large deviations in the predicted results from the experimentally determined values for some halogenated and different isomers of nitroaromatics, a more reliable correlation of the following form has been introduced [249]:

$$\Delta_{fus}H = 3.817 + 1.196a + 5.8471c - 1.382d + 7.898C_{SSP}, \tag{4.3}$$

where the C_{SSP} factor can correct the predicted results on the basis of a, c and d. For the presence of some polar groups and isomers, positive C_{SSP} values can increase the predicted $\Delta_{fus}H$ on the basis of a, c and d. In contrast, the attachment of tertiary and secondary amines as well as Ar–O– to nitroaromatic rings can result in the reverse situation being observed. The two opposite effects of the C_{SSP} can be specified according to the following conditions.

(1) Increasing effects of C_{SSP}: Increasing the number of hydroxyl (–OH) groups that are attached to an aromatic ring can increase the value for the enthalpy of melting because the strong hydrogen bonding which can result leads to a much more efficient crystal packing. Thus, the C_{SSP} value is equal to the number of –OH groups, except in the case of one hydroxyl group located between two nitro groups

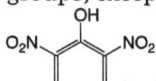

$$O_2N\diagdown\quad\overset{\text{OH}}{\diagup}\quad\diagdown NO_2 \;.$$

   (a) Or if the hydroxyl group is *ortho* to the alkyl group, in which cases the value of C_{SSP} is 0.35.

   (b) Polar groups –C(=O)–C(=O)– and –S(O)$_2$– as well as –COOH: The values of C_{SSP} are 1.0 and 1.5 for the presence of –C(=O)–C(=O)– (or –S(O)$_2$–) and –COOH groups, respectively.

   (c) Amino (–NH$_2$) groups: if amino groups are present in nitroaromatic compounds, the predicted $\Delta_{fus}H$ can be higher. If one or two amino groups are present, $C_{SSP} = 0.5$, except if there is an amino group in the *ortho* position with respect to the nitro group in mononitro derivatives. If more than two amino groups are attached to the aromatic ring, $C_{SSP} = 2.5$.

   (d) Some polar groups in the *para* position relative to the nitro group in benzene rings: If –OH and –NH$_2$ groups are in the *para* position relative to a nitro group in disubstituted or halogenated benzene rings, the corresponding values for C_{SSP} from parts (a) and (c) should be replaced by 1.5 and 1.0, respectively.

   (e) Two nitro groups: in nitroaromatic compounds which contain only two nitro groups, the values of C_{SSP} are 0.50 and 1.5 for the *ortho* (or *meta*) and other positions, respectively.

   (f) If more than one alkyl group is attached to one benzene ring (or to two benzene rings that are not directly attached to each other): $C_{SSP} = 4/n_R$ in which n_R is the number of alkyl groups. Increasing n_R can decrease the planarity of the molecule and hence the molecular attractions.

(2) Decreasing effects of C_{SSP}: The value of C_{SSP} is –1.0 if tertiary, secondary amines or Ar–O– are attached to a nitroaromatic ring. This is because the presence of these groups can decrease the planarity of the molecules, as well as reduce the interactions between local dipole moments of neighboring nitro groups.

Example 4.3. Using equation (4.3) for the following structure gives

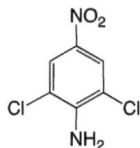

$$\Delta_{fus}H = 3.817 + 1.196a + 5.8471c - 1.382d + 7.898C_{SSP}$$
$$= 3.817 + 1.196(6) + 5.8471(2) - 1.382(2) + 7.898(1.0)$$
$$= 27.82 \, kJ/mol.$$

The predicted value is closer to the value from the experimental data (32.64 kJ/mol [142]) than equation (4.2) and group additivity method [194] are, which give values of 20.58 and 18.29 kJ/mol, respectively.

4.1.4 Nonaromatic energetic compounds containing nitramine, nitrate and nitro functional groups

A simple correlation to predict the $\Delta_{fus}H$ of acyclic and cyclic nitramines, nitrate esters and nitroaliphatic energetic compounds can be written as follows [246]:

$$\Delta_{fus}H = 16.81 + 1.896d + 4.186n_{EDNA}$$
$$+ 17.51(n_{NNO_2}^{>3,linear} - 2) - 11.52C_{-NO_2(-ONO_2)}, \tag{4.4}$$

where n_{EDNA} is the number of N,N′-(ethane-1,2-diyl)dinitramide (EDNA) moieties ($O_2NNCH_xCH_xNNO_2$) in cyclic nitramines; $n_{NNO_2}^{>3,linear}$ is the number of $-NNO_2$ groups for those acyclic linear nitramines containing more than three nitramine groups, and $C_{-NO_2(-ONO_2)}$ is a parameter. The value of $C_{-NO_2(-ONO_2)}$ is 1.0 for those compounds which contain less than four $-NO_2$ and $-ONO_2$ groups, and is $C_{-NO_2(-ONO_2)}$ is -0.75 for those compounds with four or more $-NO_2$ and $-ONO_2$ groups. The value of n_{EDNA} is zero for the simultaneous presence of $O_2NNCH_xNNO_2$ and $O_2NNCH_xCH_xNNO_2$ units in nitramine compounds (e. g. 1,3-dinitroimidazolidine (CPX)), which may be due to the lowering of the symmetry in these compounds.

Example 4.4. 1,4,5,8-Tetranitrodecahydropyrazino[2,3-b]pyrazine (TNAD) is a high performance explosive with the following molecular structure:

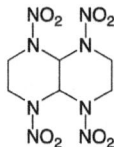

Using equation (4.4) for TNAD gives

$$\Delta_{fus}H = 16.81 + 1.896d + 4.186n_{EDNA}$$
$$+ 17.51(n_{NNO_2}^{>3,linear} - 2) - 11.52C_{-NO_2(-ONO_2)}$$
$$= 16.81 + 1.896(8) + 4.186(3) + 17.51(0) - 11.52(0)$$
$$= 44.54 \text{ kJ/mol}.$$

The measured $\Delta_{fus}H$ for TNAD is 46.4 kJ/mol [235]. The predicted $\Delta_{fus}H$ using the group additivity method is 42.66 kJ/mol [194], which shows a larger deviation from the experimentally measured value than the value obtained using equation (4.3).

4.1.5 Improved method for the reliable prediction of the enthalpy of fusion of energetic compounds

It was found that the following correlation can be used to obtain reliable predictions of the enthalpy of fusion of energetic compounds with the general formula $C_aH_bN_c(O \text{ or } S)_{d''}(halogen)_k$ [245]:

$$\Delta_{fus}H = 0.542a + 1.490b + 2.044c + 1.252d'' + 1.839k$$
$$+ 9.848\Delta H_{Inc,fus} - 11.675\Delta H_{Dec,fus}, \tag{4.5}$$

where $\Delta H_{Inc,fus}$ and $\Delta H_{Dec,fus}$ are two correcting functions which are described in the following sections.

The presence of some specific polar groups
Polar groups such as –COOH, –OH and –NH$_2$ may result in the presence of strong hydrogen bonding and consequently much more efficient packing as a result of the attractive forces. For acyclic nitamines, increasing the number of polar nitamine groups can result in enhancement of the electrostatic attractions. The contributions of $\Delta H_{Inc,fus}$ for these polar groups are:

(1) The effects of –OH and –COOH groups: The value of $\Delta H_{Inc,fus}$ is 0.4 for the presence of these functional groups, except in the case of a single hydroxyl group located

between two nitro groups , or in an *ortho* position relative to an alkyl group for which cases the value of $\Delta H_{Inc,fus} = 0.0$.

(2) The influence of –NH$_2$ groups: If one or two amino groups are present, the value of $\Delta H_{Inc,fus}$ is 0.7, except if there is an amino group in an *ortho* position relative to the nitro group for mononitro derivatives in which case $\Delta H_{Inc,fus} = 0.0$. The value of $\Delta H_{Inc,fus}$ is 2.6 if more than two amino groups are attached to an aromatic ring.

(3) The number of $-NNO_2$ groups in acyclic nitramines $n_{NNO_2}^{acycl}$: If more than three nitramine groups are present,

$$\Delta H_{Inc,fus} = n_{NNO_2}^{acycl} - 2.$$

Disubstituted nitroaromatics

Molecular interactions such as interactions between local dipole moments of neighboring atoms or groups in certain compounds can be responsible for the close proximity of molecules in the crystal. The value of $\Delta H_{Inc,fus}$ is 1.1 for dinitronaphthalene and also for disubstituted benzene derivatives in which polar groups such as $-OH$ and $-NO_2$ are located *para* to the nitro group.

Structural parameters affecting $\Delta H_{Dec,fus}$

(1) Nitroaromatics with more than one benzene ring: The value of $\Delta H_{Dec,fus}$ is equal to 0.5, except if sulfur is present between two benzene rings.
(2) Cyclic nitramines with rings which are larger than six membered rings and which contain only carbon and nitrogen atoms ($m_{cyc}^{>6}$):

$$\Delta H_{Dec,fus} = \frac{m_{cyc}^{>6} - 6}{4} + 0.5.$$

Different effects of $-NO_2$ and $-ONO_2$ groups in nonaromatics

Intermolecular interactions may be increased if a larger number of $-NO_2$ and $-ONO_2$ groups are present. Furthermore, the presence of a lower number of $-NO_2$ and $-ONO_2$ groups can result in a reduction in the molecular packing. The values of $\Delta H_{Dec,fus}$ and $\Delta H_{Inc,fus}$ are 0.4 and 1.1 if three or less $-NO_2$ (or $-ONO_2$) groups and more than three $-NO_2$ (or $-ONO_2$) groups are present, respectively.

The predicted values of $\Delta H_{Inc,fus}$ and $\Delta H_{Dec,fus}$ using the above conditions are summarized in Tables 4.2, 4.3 and 4.4.

Example 4.5. The use of equation (4.5) for the following molecular structure gives the following:

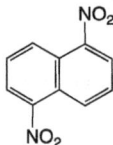

Table 4.2: Summary of the correcting function $\Delta H_{\text{Inc,fus}}$.

	Polar groups	$\Delta H_{\text{Inc,fus}}$	Condition	Exception
The presence of some specific polar groups	The effects of –OH and –COOH groups	0.4	–	One hydroxyl group between two nitro groups or *ortho* position to alkyl group
	The influence of –NH$_2$ groups	0.7	If one or two amino groups are present	If an amino group is in the *ortho* position relative to nitro group for mononitro derivatives
		2.6	If more than two amino groups are attached to an aromatic ring	–
	The number of –NNO$_2$ groups in acyclic nitramines $n_{\text{NNO}_2}^{\text{acycl}}$	$n_{\text{NNO}_2}^{\text{acycl}} - 2$	For more than three nitramine groups	–
Disubstituated nitroaromatics	–	1.1	For dinitronaphthalene or if polar groups such as –OH and –NO$_2$ are present in *para* positions to nitro group in disubstituated benzene derivatives	–

Table 4.3: Summary of the correcting function $\Delta H_{\text{Dec,fus}}$.

	$\Delta H_{\text{Dec,fus}}$	Condition	Exception
Nitroaromatics	0.5	More than one benzene ring	Sulfur is present between two benzene rings
Cyclic nitramines	$\frac{m_{\text{cyc}}^{>6} - 6}{4} + 0.5$	Rings larger than six membered rings which contain only carbon and hydrogen atoms ($m_{\text{cyc}}^{>6}$)	–

Table 4.4: Different effects of $-NO_2$ and $-ONO_2$ groups in $\Delta H_{Inc,fus}$ and $\Delta H_{Dec,fus}$.

	$\Delta H_{Inc,fus}$	$\Delta H_{Dec,fus}$	Condition
Non-aromatics	1.1	–	More than three $-NO_2$ (or $-ONO_2$) groups
	–	0.4	Three or less $-NO_2$ (or $-ONO_2$) groups

$$\Delta_{fus}H = 0.542a + 1.490b + 2.044c + 1.252d'' + 1.839k$$
$$+ 9.848\Delta H_{Inc,fus} - 11.675\Delta H_{Dec,fus}$$
$$= 0.542(10) + 1.490(6) + 2.044(2) + 1.252(4) + 1.839(0)$$
$$+ 9.848(1.1) - 11.675(0)$$
$$= 34.29 \text{ kJ/mol}.$$

The measured $\Delta_{fus}H$ for this compound is 33.03 kJ/mol [237]. The use of equation (4.3) gives a value of 33.79 kJ/mol, whereas the predicted $\Delta_{fus}H$ by the group additivity method is 17.14 kJ/mol [194], which shows a much larger deviation from the measured value.

4.1.6 A reliable method to predict the enthalpy of fusion of energetic materials

The following improved simple approach has been introduced to enable prediction of the values of $\Delta_{fus}H$ in large classes of energetic compounds – including polynitro arenes, polynitro heteroarenes, acyclic and cyclic nitramines, nitrate esters, nitroaliphatics, cyclic and acyclic peroxides, as well as nitrogen rich compounds [251]:

$$\Delta_{fus}H = 0.9781(\Delta_{fus}H)_{add} + 7.567(\Delta_{fus}H)^{Inc}_{nonadd} - 8.784(\Delta_{fus}H)^{Dec}_{nonadd}, \quad (4.6)$$

where

$$(\Delta_{fus}H)_{add} = 0.6047a + 0.6211b + 2.750c + 1.424d'' + 3.048k. \quad (4.7)$$

$(\Delta_{fus}H)^{Inc}_{nonadd}$ and $(\Delta_{fus}H)^{Dec}_{nonadd}$ are also nonadditive contributions corresponding to the increasing and decreasing effects of specific groups. The presence of polar groups results in an increase in the intermolecular attractions of a molecule with its neighboring molecules. The values of $(\Delta_{fus}H)^{Inc}_{nonadd}$ and $(\Delta_{fus}H)^{Dec}_{nonadd}$ are described in the following sections.

$(\Delta_{fus}H)^{Inc}_{nonadd}$

(1) $-OH$ and $-COOH$ groups in aromatic compounds: The value of $(\Delta_{fus}H)^{Inc}_{nonadd}$ is equal to 0.7, except if an $-OH$ group is located between two nitro groups in an aromatic structure for which the value of $(\Delta_{fus}H)^{Inc}_{nonadd}$ is zero.

(2) $-NH_2$ group: The value of $(\Delta_{fus}H)^{Inc}_{nonadd}$ is 0.5 if one or two amino groups are present in aromatic or non-aromatic compounds. The value of $(\Delta_{fus}H)^{Inc}_{nonadd}$ equals 2.6 if more than two amino groups are attached to aromatic rings. For mononitro derivatives, if the nitro group is located in the *ortho* position with respect to the amino group, $(\Delta_{fus}H)^{Inc}_{nonadd}$ is zero.

(3) >NH group: if the >NH fragment is present, $(\Delta_{fus}H)^{Inc}_{nonadd}$ equals 0.5.

(4) $>N-NO_2$ group in acyclic nitramines: if more than three $>N-NO_2$ groups are present, $(\Delta_{fus}H)^{Inc}_{nonadd}$ is $(n^{acyclic}_{NNO_2} - 2)$ where $n^{acyclic}_{NNO_2}$ is the number of $>N-NO_2$ groups in acyclic nitramines.

(5) $-N(C=O)N-$ group: The value of $(\Delta_{fus}H)^{Inc}_{nonadd}$ is 1.5.

$(\Delta_{fus}H)^{Dec}_{nonadd}$

(1) Ar–X–Ar nitroaromatics: The value of $(\Delta_{fus}H)^{Dec}_{nonadd}$ is 0.5, except for Ar–S–Ar structures for which it is zero.

(2) Cyclic nitramines:

$$(\Delta_{fus}H)^{Dec}_{nonadd} = \frac{m_{cyc} - 6}{4} + 0.5$$

for cyclic nitramines in which the ring(s) are larger than six-membered rings and which contain only carbon and nitrogen atoms ($m_{cyc} > 6$).

(3) $-NO_2$ group in non-aromatic compounds: The value of $(\Delta_{fus}H)^{Dec}_{nonadd}$ is 1.0 for the presence of nitro groups in acyclic and cyclic alkanes.

(4) $-N-N=O$ group: The value of $(\Delta_{fus}H)^{Dec}_{nonadd}$ is 2.0 if the nitroso group is present.

Table 4.5 summarizes some functional groups and molecular fragments which can be used to determine the values of $(\Delta_{fus}H)^{Inc}_{nonadd}$ and $(\Delta_{fus}H)^{Dec}_{nonadd}$.

Example 4.6. The use of equations (4.6) and (4.7) for different classes of energetic compounds, including acyclic nitramines, cyclic nitramines, nitrate esters, polynitro arenes, polynitro heteroarenes, nitroaliphatics, nitroaromatics, cyclic peroxides, acyclic peroxides and nitrogen rich compounds, is given in Table 4.6.

4.2 Different methods to predict the entropy of fusion

Organic molecules with high symmetry may have lower entropy of fusion and a higher melting point as compared to their nonsymmetrical isomers. Their molecules have a higher probability of being in the right orientation to form the crystal. It was shown that some of the entropy determining parameters are not group additive [196, 259, 260]. There are more translational, rotational, and conformational constraints on molecules in the solid state than the liquid state. The entropy of fusion ($\Delta_{fus}S$) is an important property for predicting the melting point and solubility of organic compounds [253]. The value of $\Delta_{fus}S$ based on Walden's rule is constant with

Table 4.5: Correcting functions which are used for the presence of different molecular fragments.

Molecular moieties	Effect on predicted $\Delta_{fus}H$		Comment
	$(\Delta_{fus}H)^{Inc}_{nonadd}$	$(\Delta_{fus}H)^{Dec}_{nonadd}$	
−OH and −COOH groups	0.7		In aromatic compounds
	0.0		In O_2N–C(OH)=C–NO_2 fragments
−NH_2 group	0.5		For one or two −NH_2 groups
	2.6		For more than two −NH_2 groups
	0.0		In (aromatic with NO_2 and NH_2) fragments
>NH group	0.5		–
>NNO_2 group	$n_{NNO_2} - 2$		In acyclic nitramines
−N(C=O)N− fragment	1.5		–
Ar−X−Ar nitroaromatics		0.5	–
		0.0	where X is S
Cyclic nitramines		$\frac{m_{cyc}-6}{4} + 0.5$	Rings which are larger than six membered rings
−NO_2 group		1.0	Nitroalkanes
−N−N=O group		2.0	–

a value of $56.5\,\mathrm{J\,K^{-1}\,mol^{-1}}$ [261] for aromatic compounds with little variation. The effect of molecular rotational symmetry (σ) can be incorporated into Walden's rule as follows [262]:

$$\Delta_{fus}S = 56.5 - R \ln \sigma. \tag{4.8}$$

The parameter σ is the rotational degeneracy of the molecule. It is the number of positions a molecule can be rotated into while maintaining the same atomic orientation of the original position. Figure 4.1 shows the symmetry number for some typical compounds. The value of σ is based upon rotation because it is the only symmetry operation that can physically be performed on a molecule. Thus, it does not enumerate mirror planes or operations that cannot be physically performed. Hydrogen atoms in OH, CH_3, and NH_2 groups as well as halogens of tri-homohalogenated carbons like trichloromethyl are treated as being radially symmetrical because they can be assumed to be freely and rapidly rotating. Carboxylic acids and nitro groups have

Table 4.6: Several examples for the use of equations (4.6) and (4.7) in some classes of energetic compounds.

Class of compound	Molecular structure	The calculated $\Delta_{fus}H$
Acyclic nitramine	$H_2C{=}NNO_2$	$\Delta_{fus}H = 0.9781[0.6047(1) + 0.6211(2)$ $+ 2.750(2) + 1.424(2) + 3.048(0)]$ $+ 7.567(0) - 8.784(0) = 9.97\,kJ/mol$
		$\Delta_{fus}H = 0.9781[0.6047(2) + 0.6211(6)$ $+ 2.750(4) + 1.424(4) + 3.048(0)]$ $+ 7.567(0.5) - 8.784(0)$ $= 24.94\,kJ/mol$
Cyclic nitramine		$\Delta_{fus}H = 0.9781[0.6047(6) + 0.6211(6)$ $+ 2.750(12) + 1.424(12)$ $+ 3.048(0)] + 7.567(0) - 8.784(0)$ $= 56.18\,kJ/mol$
		$\Delta_{fus}H = 0.9781[0.6047(4) + 0.6211(8)$ $+ 2.750(8) + 1.424(8) + 3.048(0)]$ $+ 7.567(0) - 8.784(1)$ $= 31.10\,kJ/mol$
Nitrate ester		$\Delta_{fus}H = 0.9781[0.6047(4) + 0.6211(8)$ $+ 2.750(2) + 1.424(7) + 3.048(0)]$ $+ 7.567(0) - 8.784(0)$ $= 22.35\,kJ/mol$
	$O_2N{-}O{-}CH_2\,CH_2{-}O{-}NO_2$ $O_2N{-}O{-}CH_2\,CH_2{-}O{-}NO_2$	$\Delta_{fus}H = 0.9781[0.6047(5) + 0.6211(8)$ $+ 2.750(4) + 1.424(12) + 3.048(0)]$ $+ 7.567(0) - 8.784(0)$ $= 35.29\,kJ/mol$
Polynitro arene		$\Delta_{fus}H = 0.9781[0.6047(14) + 0.6211(8)$ $+ 2.750(6) + 1.424(12) + 3.048(0)]$ $+ 7.567(0) - 8.784(0.5)$ $= 45.99 - 4.39 = 41.60\,kJ/mol$
		$\Delta_{fus}H = 0.9781[0.6047(6) + 0.6211(5)$ $+ 2.750(5) + 1.424(6) + 3.048(0)]$ $+ 7.567(0.5) - 8.784(0)$ $= 32.17\,kJ/mol$
Polynitro heteroarene		$\Delta_{fus}H = 0.9781[0.6047(5) + 0.6211(2)$ $+ 2.750(4) + 1.424(6) + 3.048(0)]$ $+ 7.567(0) - 8.784(0) = 23.29\,kJ/mol$

Table 4.6 (continued)

Class of compound	Molecular structure	The calculated $\Delta_{fus}H$
		$\Delta_{fus}H = 0.9781[0.6047(6) + 0.6211(2) + 2.750(4)$ $+ 1.424(6) + 3.048(0)] + 7.567(0) - 8.784(0)$ $= 23.88$ kJ/mol
Nitroaliphatic	$H_3C{-\!\!-}NO_2$	$\Delta_{fus}H = 0.9781[0.6047(1) + 0.6211(3) + 2.750(1)$ $+ 1.424(2) + 3.048(0)] + 7.567(0) - 8.784(0)$ $= 7.89$ kJ/mol
		$\Delta_{fus}H = 0.9781[0.6047(2) + 0.6211(3) + 2.750(3)$ $+ 1.424(7) + 3.048(0)]$ $+ 7.567(0) - 8.784(1) = 12.04$ kJ/mol
Nitroaromatic		$\Delta_{fus}H = 0.9781[0.6047(6) + 0.6211(4) + 2.750(1)$ $+ 1.424(2) + 3.048(1)]$ $+ 7.567(0) - 8.784(0) = 14.43$ kJ/mol
		$\Delta_{fus}H = 0.9781[0.6047(7) + 0.6211(5) + 2.750(1)$ $+ 1.424(4) + 3.048(0)]$ $+ 7.567(0.7) - 8.784(0) = 20.73$ kJ/mol
Cyclic peroxide		$\Delta_{fus}H = 0.9781[0.6047(6) + 0.6211(12) + 2.750(0)$ $+ 1.424(4) + 3.048(0)]$ $+ 7.567(0) - 8.784(0) = 16.41$ kJ/mol
		$\Delta_{fus}H = 0.9781[0.6047(9) + 0.6211(18) + 2.750(0)$ $+ 1.424(6) + 3.048(0)]$ $+ 7.567(0) - 8.784(0) = 24.61$ kJ/mol
Acyclic peroxide		$\Delta_{fus}H = 0.9781[0.6047(14) + 0.6211(10)$ $+ 2.750(0) + 1.424(4) + 3.048(0)]$ $+ 7.567(0) - 8.784(0) = 19.93$ kJ/mol
Nitrogen-rich compound		$\Delta_{fus}H = 0.9781[0.6047(3) + 0.6211(6)$ $+ 2.750(4) + 1.424(0) + 3.048(0)]$ $+ 7.567(0) - 8.784(0) = 16.18$ kJ/mol
		$\Delta_{fus}H = 0.9781[0.6047(9) + 0.6211(5)$ $+ 2.750(4) + 1.424(0) + 3.048(3)]$ $+ 7.567(0.5) - 8.784(0.5) = 27.46$ kJ/mol

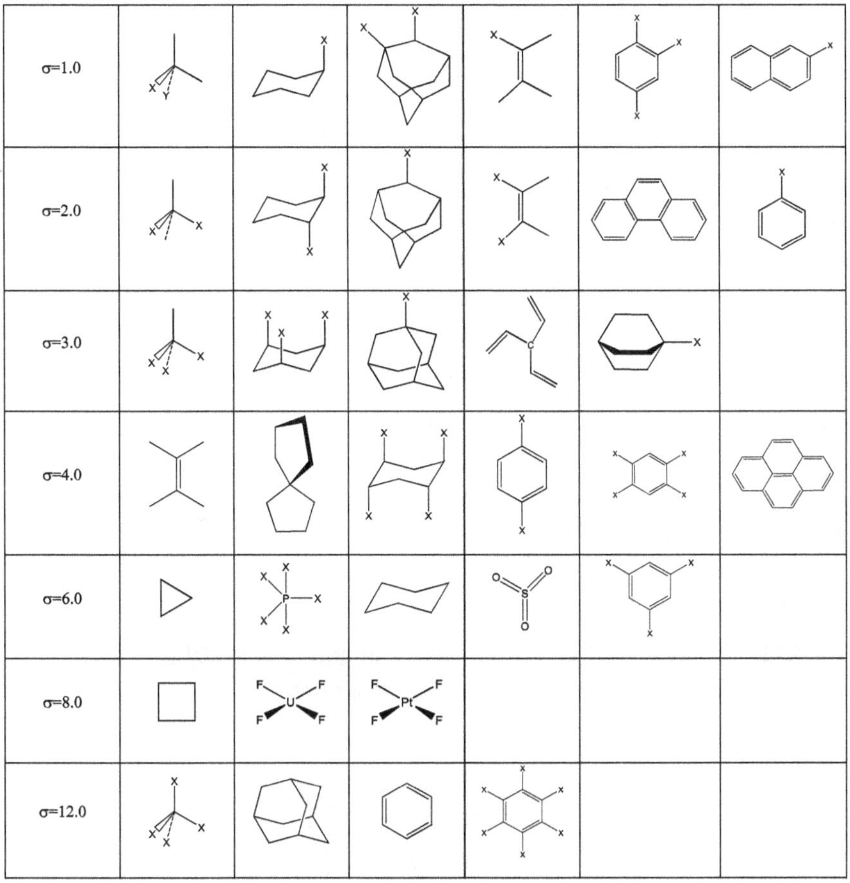

Figure 4.1: The value of σ for some representative compounds.

bilateral symmetry. The molecules with geometry of cones, cylinders, and spheres have infinite symmetry for which σ is taken 20 to reflect their higher symmetry. Thus, these geometries are not considered as equal to unity as in chemistry. Rather they have σ = 20 because they are more symmetrical than benzene or neopentane as well as containing at least one infinite rotational axis. The value of σ is equal to 1.0 for all nonsymmetrical molecules.

The parameter σ′ can be considered for those molecules with a slight structural difference, which causes an otherwise symmetrical molecule to be asymmetrical. The packing arrangement of this molecule is quite the same as that of its symmetrical homomorphy, i. e. pseudo-symmetrical molecules have a higher probability of being in the right arrangement required for a crystal lattice than nonsymmetrical molecules. Thus, halogens and methyl group are treated as being pseudosymmetrical atoms/group, e. g. p-bromochlorobenzene is a pseudosymmetrical molecule (σ′ = 2) as compared to its homomorph (p-dichlorobenzene or p-dibromobenzene, σ = 2).

Equation (4.8) can be modified to include the flexibility number (ϕ) as [263]:

$$\Delta_{fus}S = 50 - R \ln \sigma + R \ln \phi. \tag{4.9}$$

The parameter ϕ shows the internal conformational freedom of molecules. Flexible molecules, as compared to rigid molecules, tend to have a greater entropy change during melting. This parameter is calculated by an *ad hoc* expression uniting flexible segments as:

$$\phi = 0.3ROT + LINSP3 + 0.5(BRSP3 + SP2 + RING) - 1, \tag{4.10}$$

where *LINSP3* is the number of nonring, nonterminal, and nonbranched sp^3 atoms; ROT is the extra entropy produced by freely rotating sp^3 atoms and is calculated as *ROT = LINSP3-4* (if *LINSP3* > 4 otherwise *ROT* = 0); *BRSP3* is the total number of nonring, nonterminal, and branched sp^3 atoms; *SP2* is the number of nonring, and nonterminal sp^2 atoms; RING is the number of single, fused, or conjugated ring systems. If the calculated value of ϕ by equation (4.3) is less than zero, the value of ϕ should be taken zero.

Example 4.7. The values of different parameters of equation (4.10) are specified for 1-methoxy-2,4-dinitrobenzene with the following structure:

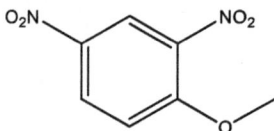

$$\phi = 0.3ROT + LINSP3 + 0.5(BRSP3 + SP2 + RING) - 1,$$

LINSP3 = 1 (the oxygen atom of methoxy group), *BRSP3* = 0, *SP2* = 2 (two nitrogen atoms of nitro groups), *RING* = 1 (one benzene ring), and *ROT* = 0 because *LINSP3* < 4. The use of these parameters in equation (4.10) gives ϕ = 1.5.

To predict $\Delta_{fus}S$ of nitroaromatic compounds, a suitable correlation based on the elemental composition has been used [259]. The reliability of this method is higher than that of the method of Jain and Yalkowsky [194], which is based on equation (4.8) for 61 nitroaromatic compounds. This method has been restricted to nitroarmatic compounds because it may give values which show large deviation from the expected values for some of the other classes of energetic compounds. For nitroaromatics, Evan and Yallkowsky [196] have improved the reliability of the method of Jain and Yalkowsky [194] by introducing the molecular eccentricity (ε):

$$\Delta_{fus}S = 50 - R \ln \sigma + R \ln \phi + R \ln \varepsilon. \tag{4.11}$$

Yalkowsky and coworkers [202, 264] have improved various parameters to obtain the entropy of fusion for over 2000 organic compounds with a reasonable approximation as:

$$\Delta_{fus}S = 44.98 - 8.93 \log \sigma - 2.17 \log \sigma' + 11.36 \log \varepsilon_{ar} + 8.26 \log \varepsilon_{al} + 5.91\phi, \quad (4.12)$$

where σ is symmetry, σ' is pseudosymmetry, ε_{ar} and ε_{al} are aromatic and aliphatic eccentricity, respectively, and ϕ is flexibility.

The parameters ε_{ar} and ε_{al} are related to crystals of flat eccentric molecules, which have relatively less than average free volume due to their efficient packing. Flat eccentric molecules require more space to attain their free rotation as compared to spherical molecules of the same molecular weight. These parameters can be obtained by the number of atoms in aliphatic rings as well as the number of atoms in aromatic rings or directly connected to them or part of a conjugated system, e. g. ε_{ar} is equal to 16 for 2,4,6-trinitrotoluene due to the existence of three nitro conjugated system (= 3 × 3), six carbon atoms of the benzene ring and one carbon atom of the methyl group. The values of ε_{ar} and ε_{al} equal to one for all nonring compounds.

Since the free energy of an organic compound at its melting point is equal to zero, the normal melting point is the temperature at which the solid and liquid are at equilibrium at one-atmosphere pressure. Thus, the use of equations (4.11) and (4.12) can give the normal melting point of an organic compound as:

$$T_m = \frac{\Delta_{fus}H}{\Delta_{fus}S} = \frac{\sum n_i H_{fus,i}}{44.98 - 8.93 \log \sigma - 2.17 \log \sigma' + 11.36 \log \varepsilon_{ar} + 8.26 \log \varepsilon_{al} + 5.91\phi}.$$
$$(4.13)$$

Example 4.8. 2,4,6-Trinitro-N-methyl-aniline has the following structure:

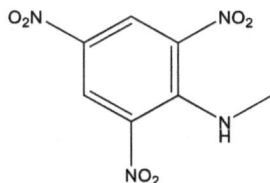

The value of $\Delta_{fus}H$ is calculated by equation (4.11) and Table 4.1 as follows:

$$\Delta_{fus}H = \sum n_i H_{fus,i} = 4C_{ar} + 2CH_{ar} + 3YNO_2 + YNH + XCH_3$$
$$= 4 \times 1.777 + 2 \times 1.235 + 3 \times 4.584 + 3.799 + 0.701 = 27.83 \text{ kJ mol}^{-1}$$

The measured value of $\Delta_{fus}H$ for this compound is 25.9 kJ mol^{-1} [202].

The value of $\Delta_{fus}S$ is calculated by equation (4.12) as follows:

$$\Delta_{fus}S = 44.98 - 8.93 \log \sigma - 2.17 \log \sigma' + 11.36 \log \varepsilon_{ar} + 8.26 \log \varepsilon_{al} + 5.91\phi$$
$$= 44.98 - 8.93 \log 1 - 2.17 \log 1 + 11.36 \log 16 + 8.26 \log 1 + 5.91 \times 2.0$$
$$= 70.5 \text{ J mol}^{-1}\text{K}^{-1}$$

The experimental value of $\Delta_{fus}S$ is 64.2 J mol^{-1} K^{-1} [202].

A general new method for predicting the entropy of fusion for various types of energetic compounds – including polynitro arene, acyclic and cyclic nitramine, nitrate esters and nitroaliphatic compounds – has been introduced as additive ($\Delta_{fus}S_{add}$) and nonadditive ($\Delta_{fus}S_{nonadd}$) functions [260]:

$$\Delta_{fus}S = \Delta_{fus}S_{add} - 23.86\Delta_{fus}S_{nonadd}, \tag{4.14}$$

where

$$\Delta_{fus}S_{add} = 39.99 + 5.88c + 1.22d. \tag{4.15}$$

The values of $\Delta_{fus}S_{nonadd}$ can be specified as follows.
(1) Ar–N(NO$_2$)– and –R–OH molecular moieties: The values of $\Delta_{fus}S_{nonadd}$ are 1.0 and 1.7 respectively, for the presence of these molecular fragmnents.
(2) 1,3,5-Trinitrobenzene derivatives: if 1,3,5-trinitrobenzene, 2,2′,4,4′,6,6′-hexanitro-biphenyl or two rings of 1,3,5-trinitrobenzene in the form

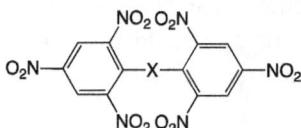

are present, the value of $\Delta_{fus}S_{nonadd}$ is 1.0 if X is a single atom or unsaturated molecular fragment (except –N=N– connecting two 1,3,5-trinitrobenzene rings).
(3) Cyclic nitramines containing more than three –NNO$_2$ groups: $\Delta_{fus}S_{nonadd} = 1.5$.

Example 4.9. The use of equations (4.14) and (4.15) for the following molecular structure

gives the following value:

$$\begin{aligned}
\Delta_{fus}S_{add} &= 39.99 + 5.88c + 1.22d \\
&= 39.99 + 5.88(7) + 1.22(12) \\
&= 95.79 \, \text{J K}^{-1} \text{mol}^{-1}, \\
\Delta_{fus}S &= \Delta_{fus}S_{add} - 23.86\Delta_{fus}S_{nonadd} \\
&= 95.79 - 23.86(1) \\
&= 71.93 \, \text{J K}^{-1} \text{mol}^{-1}.
\end{aligned}$$

The measured $\Delta_{fus}S$ for this compound is 72.8 J K^{-1} mol^{-1} [54]. The predicted $\Delta_{fus}S$ using equation (4.9) is 55.9 J K^{-1} mol^{-1}, which deviates considerably from the measured value.

4.3 Summary

Different group additivity and empirical methods have been reviewed which can be used to predict the enthalpy and entropy of fusion. Equation (4.1) and Table 4.1 provide a suitable group additivity method for prediction of the enthalpy of fusion of organic compounds containing organic materials with energetic groups. Among the different methods for predicting the enthalpy of fusion, equations (4.6) and (4.7) provide a simple pathway for large classes of energetic compounds – including polynitro arene, polynitro heteroarene, acyclic and cyclic nitramines, nitrate esters, nitroaliphatic, cyclic and acyclic peroxides, as well as nitrogen-rich compounds. Equation (4.12) gives a suitable approaction for estimation of the entropy of fusion of organic compounds where it can be applied for organic energetic materials. Equations (4.14) and (4.15) also provide a new general method for predicting the entropy of fusion for various types of energetic compounds including polynitro arenes, acyclic and cyclic nitramines, nitrate esters and nitroaliphatic compounds.

5 Heat of sublimation

Sublimation properties can help to understand the coarsening process which occurs [265, 266]. Knowledge of the heat of sublimation and the vapor pressure of organic explosives can also help to enable the design of new technology for the detection of explosive particles from concealed devices [267]. The sublimation properties of organic explosives have long-term effects on soil, water and air. Since secondary explosives are widely used, understanding their sublimation properties is necessary in order to be able to study their toxic effects in the environment during storage [268].

The heat – or enthalpy – of sublimation is the best parameter to characterize the strength of intermolecular interactions within a crystal. The gas phase heat of formation ($\Delta_f H(g)$) and the heat of sublimation ($\Delta_{sub}H$) can be used to evaluate the solid phase heats of formation ($\Delta_f H(s)$) as follows [269]:

$$\Delta_f H(s) = \Delta_f H(g) - \Delta_{sub}H. \tag{5.1}$$

The $\Delta_f H(s)$ value of an energetic compound is important in order to enable the prediction of its performance using various computer codes such as CHEETAH [270], ISPBKW [271], LOTUSES [272], EDPHT [87], and EMDB [69]. It can also be calculated by combining the predicted heat of sublimation and gas phase heat of formation according to equation (5.1).

The value of $\Delta_{sub}H$ for a specific compound can be considered to be the sum of its heat of fusion and its heat of vaporization, even if the liquid cannot exist at the pressure and temperature in question. Since the sublimation pressure at the melting point is only rarely known, it is difficult to use the Clausius–Clapeyron equation to obtain the heat of sublimation from the vapor pressure data [253].

There are a variety of approaches which can be used to predict the gas phase heats of formation of energetic compounds [91, 92, 113, 134, 253, 258, 273], but few methods have been reported to predict the heat of sublimation of energetic compounds. Moreover, experimental data for the heats of sublimation of energetic compounds are rare because these values have not yet been published for many energetic materials. In this chapter, some methods for the prediction of $\Delta_{sub}H$ for several classes of energetic compounds will be reviewed.

https://doi.org/10.1515/9783110740158-005

5.1 Group additivity method for prediction of the heat of sublimation

Mathieu [94] generated a model using 35 group contributions from a training set containing 814 compounds. Naef and Acree Jr. [274] introduced a suitable group additivity method for predicting the heat of sublimation of a wide range of organic compounds containing energetic materials. The general form of equation (5.2) can be used for this purpose:

$$\Delta_{sub}H = 21.03 + \sum_i a_i A_i + \sum_j g_j G_j, \qquad (5.2)$$

where a_i and g_j are the numbers of the ith atom group A_i and the special group G_j, respectively. For the atom groups, each group consists of a central atom (the backbone atom) and its immediate neighbor atoms. The central atom is bound to at least two other atoms. It is characterized by its atom name, its atom type being defined by either its orbital hybridization or bond type or its number of bonds, where required for distinction, and by its charge, if not zero. A term is used to collect the neighbour atoms, which are in the order H > B > C > N > O > S > P > Si > F > Cl > Br > I. The bond type of the neighbor bond with the backbone atom (if not single) encompasses its atom name and its number of occurrences (if >1). The symbol J is used instead of I because of the better readability of a neighbor term containing iodine. For the nonzero total net charge of the neighbor atoms is nonzero, the charge "(+)" or "(−)" is appended to the neighbor term. The atom type "N sp3" is used for N with three single bonds. The atom types "O" and "S" are used for O and S with two single bonds, respectively. If neighbor atoms are part of a conjugated moiety, the terms "(pi)", "(2pi)" or "(3pi)" supplement the neighbor term. The increased strength of a group's bonds in this situation is due to the π-orbital conjugation of the backbone atom's lone-pair electrons with conjugated neighbor moieties. Table 5.1 shows some examples for the backbone atom type (the boldface) and the term for its neighbors. Table 5.2 also gives atom groups and their contributions (in kJ/mol) for equation (5.2).

The use of equation (5.2) does not reflect any knowledge about the molecules' three-dimensional structure. It also depends on structural peculiarities such as buttressing effects, ring strains, gauche bond interactions, or internal hydrogen bonds.

Example 5.1. The use of equation (5.2) for (a) 1,3,5-trinitrobenzene, (b) hexanitroethane, and (c) 1,3-diamino-2,4,6-trinitrobenzene gives:

(a) According to Tables 5.1 and 5.2, 1,3,5-trinitrobenzene has three (Atomic type = C aromatic, Neighbours = H:C2, and Contribution = 5.36 kJ/mol), three (Atomic type = C aromatic, Neighbours = :C2N(+), and Contribution = 28.67 kJ/mol), and three (Atomic type = N(+) sp2, Neighbours = CO=O(−), and

Table 5.1: Group examples and their meaning.

Atom Type	Neighbors	Meaning	Atom Type	Neighbors	Meaning
C sp3	H3C	C–CH3	N sp3	H2C	C–NH2
C sp3	H3N	N–CH3	N sp3	H2C(pi)	C–N*H2
C sp3	H2C2	C–CH2–C	N sp3	C2N(2pi)	C–N*(N)–C
C sp3	H2CO	C–CH2–O	N sp2	H=C	C=NH
C sp3	HC3	C–CH(C)–C	N sp2	C=N	N=N–C
C sp3	HC2Cl	C–CH(Cl)–C	N sp2	=CO	C=N–O
C sp3	HCO2	C–CH(O)–O	N(+) sp3	H3C	C–NH$_3^+$
C sp3	C3N	C–C(C)2–N	N(+) sp3	H2C2	C–NH$_2^+$–C
C sp3	C2F2	C–CF2–C	N(+) sp2	CO=O(–)	O=N$^+$(O$^-$)–C
C sp2	H2=C	C=CH2	N aromatic	:C2	C:N:C
C sp2	HC=C	C=CH–C	N(+) sp	=N2(–)	N=N$^+$=N$^{(-)}$
C sp2	HC=N	N=CH–C	O	HC	C–OH
C sp2	H=CN	C=CH–N	O	HC(pi)	C–O*H^c
C sp2	HN=O	O=CH–N	O	Si2	Si–O–Si
C sp2	C2=O	O=C(C)–C	P3	C3	C–PC)–C
C sp2	C=CN	C=C(C)–N	P4	CO2=O	O=PO2)–C
C sp2	=CNO	C=C(N)–O	P4	N2O=O	O=PO)(N)–N
C sp2	N=NO	N=C(N)–O	S2	HC(pi)	C–S*H
C sp2	NO=O	O=C(N)–O	S2	CS	C–S–S
C aromatic	H:C2^a	C:CH:C	S4	CO=O2	C–S(=O)2–O
C aromatic	H:C:N	C:CH:N	S4	O2=O	O–S(=O)–O
C aromatic	:CN:N	C:C(N):N	Si	C2Cl2	C–SiCl$_2$–C
C sp	H#C^b	C#CH	Si	OCl3	O–SiCl$_3$
C sp	C#N	N#C–C			
C sp	#CN	C#C–N			
C sp	=C2	C=C=C			
C sp	=C=O	C=C=O			

^{a}The symbol ":" represents an aromatic bond.
^{b}The symbol "#" gives a triple bond.
^{c}The symbol "*" shows lone-pair electrons form π-orbital conjugated bonds with neighbor atoms.

Contribution = −4.38 kJ/mol).

$$\Delta_{sub}H = 21.03 + \sum_i a_i A_i + \sum_j g_j G_j$$

$$= 21.03 + 3(5.36) + 3(28.67) + 3(-4.38) = 109.98 \text{ kJ/mol}$$

(b) According to Tables 5.1 and 5.2, hexanitroethane has two (Atomic type = C sp3, Neighbours = CN3(+), and Contribution = 43.89 kJ/mol), and six (Atomic type = N(+) sp2, Neighbours = CO=O(−), and Contribution = −4.38 kJ/mol).

$$\Delta_{sub}H = 21.03 + \sum_i a_i A_i + \sum_j g_j G_j$$

$$= 21.03 + 2(43.89) + 6(-4.38) = 82.53 \text{ kJ/mol}$$

Table 5.2: Atom groups and their contributions (in kJ/mol) for equation (5.2).

Atom Type	Neighbors	Contribution
B	C3	65.82
C sp3	H3C	5.99
C sp3	H3N	26.96
C sp3	H3N(+)	98.98
C sp3	H3O	28.51
C sp3	H3S	30.06
C sp3	H2C2	6.88
C sp3	H2CN	21.98
C sp3	H2CN(+)	27.46
C sp3	H2CO	29.62
C sp3	H2CS	23.29
C sp3	H2CF	15.91
C sp3	H2CCl	17.59
C sp3	H2CBr	22.76
C sp3	H2CJ	21.83
C sp3	H2N2	43.95
C sp3	H2NCl	36.29
C sp3	H2O2	53.35
C sp3	H2OS	54.78
C sp3	H2S2	47.45
C sp3	HBC2	36.17
C sp3	HC3	2.28
C sp3	HC2N	14.28
C sp3	HC2N(+)	21.01
C sp3	HC2O	24.27
C sp3	HC2S	17.59
C sp3	HC2F	5.18
C sp3	HC2Cl	11.49
C sp3	HC2Br	0.95
C sp3	HCN2	39.48
C sp3	HCN2(+)	39.93
C sp3	HCNO	34.73
C sp3	HCNS	20.56
C sp3	HCO2	39.96
C sp3	HCF2	0.19
C sp3	HCCl2	15.78
C sp3	HN3(+)	37.31
C sp3	HO3	72.23
C sp3	C4	4.25
C sp3	C3N	5.87
C sp3	C3N(+)	18.44
C sp3	C3O	15.18
C sp3	C3S	6.40
C sp3	C3F	1.89
C sp3	C3Cl	8.06

Table 5.2 (continued)

Atom Type	Neighbors	Contribution
C sp3	C3Br	2.34
C sp3	C2N2(+)	34.78
C sp3	C2O2	39.73
C sp3	C2S2	37.28
C sp3	C2F2	7.07
C sp3	CN3(+)	43.89
C sp3	CN2F(+)	25.98
C sp3	CO3	57.42
C sp3	CF3	4.71
C sp3	CCl3	16.10
C sp3	N3F(+)	44.00
C sp3	O4	73.43
C sp2	H2=C	7.97
C sp2	HC=C	5.10
C sp2	HC=N	35.49
C sp2	HC=N(+)	72.64
C sp2	H=CN	32.79
C sp2	HC=O	20.74
C sp2	H=CO	16.89
C sp2	H=CS	15.22
C sp2	HN=N	55.52
C sp2	HN=O	35.41
C sp2	H=NO	40.91
C sp2	H=NS	33.85
C sp2	C2=C	3.91
C sp2	C2=N	30.47
C sp2	C2=N(+)	13.76
C sp2	C=CN	26.81
C sp2	C=CN(+)	41.65
C sp2	C2=O	15.10
C sp2	C=CO	22.08
C sp2	C2=S	18.21
C sp2	C=CS	15.64
C sp2	C=CF	16.81
C sp2	C=CCl	11.02
C sp2	C=CBr	34.06
C sp2	C=CJ	32.46
C sp2	=CN2	64.94
C sp2	=CN2(+)	60.65
C sp2	CN=N	54.51
C sp2	CN=N(+)	44.16
C sp2	CN=O	39.66
sp2	C=NO	42.74
C sp2	CN=S	39.85
C sp2	C=NS	34.89

Table 5.2 (continued)

Atom Type	Neighbors	Contribution
C sp2	=CNS(+)	41.29
C sp2	=CNCl	38.14
C sp2	CO=O	34.06
C sp2	CO=O(−)	80.89
C sp2	C=OCl	29.03
C sp2	CS=S	56.97
C sp2	N2=N	80.72
C sp2	N2=N(+)	65.95
C sp2	N2=O	59.57
C sp2	N2=S	66.62
C sp2	N=NS	51.62
C sp2	NO=O	52.79
C sp2	=NO2	61.12
C sp2	N=OS	48.27
C sp2	NO=S	58.04
C sp2	=NOS	52.75
C sp2	NS=S	60.83
C sp2	=NS2	64.37
C sp2	O2=O	41.40
C sp2	=OS2	41.22
C sp2	OS=S	73.06
C sp2	S2=S	49.39
C aromatic	H:C2	5.36
C aromatic	H:C:N	18.20
C aromatic	H:C:N(+)	28.26
C aromatic	H:N2	23.27
C aromatic	B:C2	25.04
C aromatic	:C3	5.51
C aromatic	C:C2	3.12
C aromatic	C:C:N	11.10
C aromatic	C:C:N(+)	16.04
C aromatic	:C2N	22.21
C aromatic	:C2N(+)	28.67
C aromatic	:C2:N	17.03
C aromatic	:C2:N(+)	18.05
C aromatic	:C2O	20.46
C aromatic	:C2P	1.63
C aromatic	:C2S	16.31
C aromatic	:C2F	4.45
C aromatic	:C2Cl	12.48
C aromatic	:C2Br	14.66
C aromatic	:C2J	20.68
C aromatic	:C2Si	4.80
C aromatic	C:N2	28.80
C aromatic	:CN:N	29.72

Table 5.2 (continued)

Atom Type	Neighbors	Contribution
C aromatic	:CN:N(+)	33.74
C aromatic	:C:NO	41.44
C aromatic	:C:NO(+)	33.50
C aromatic	:C:NCl	21.70
C aromatic	:C:NBr	31.31
C aromatic	N:N2	43.11
C aromatic	:N2O	39.92
C aromatic	:N2S	36.08
C aromatic	:N2Cl	35.90
C sp	=C2	6.39
C sp	C#C	3.24
C sp	C#N	16.49
C sp	C#N(+)	11.33
C sp	#CS	28.03
C sp	N#N	47.80
C sp	#NP	12.53
N sp3	H2C	5.03
N sp3	H2C(pi)	6.38
N sp3	H2N	17.97
N sp3	H2S	41.98
N sp3	HC2	23.83
N sp3	HC2(pi)	13.51
N sp3	HC2(2pi)	20.10
N sp3	HCN	0.15
N sp3	HCN(pi)	6.71
N sp3	HCN(2pi)	6.84
N sp3	HCS(pi)	15.10
N sp3	C3	51.07
N sp3	C3(pi)	53.90
N sp3	C3(2pi)	60.80
N sp3	C3(3pi)	61.26
N sp3	C2N(pi)	7.05
N sp3	C2N(+)(pi)	5.52
N sp3	C2N(2pi)	36.36
N sp3	C2N(+)(2pi)	20.13
N sp3	C2N(3pi)	54.74
N sp3	C2S	49.13
N sp3	C2F(2pi)	64.78
N sp3	CN2(pi)	30.74
N sp3	CN2(2pi)	49.40
N sp3	CN2(+)(2pi)	3.72
N sp3	CNF(2pi)	34.74
N sp2	C=C	32.77
N sp2	C=N	4.54
N sp2	C=N(+)	15.43

Table 5.2 (continued)

Atom Type	Neighbors	Contribution
N sp2	=CN	4.63
N sp2	=CN(+)	36.68
N sp2	C=O	12.04
N sp2	C=P	49.18
N sp2	=CO	16.24
N sp2	=CS	26.78
N sp2	N=N	12.19
N sp2	N=O	0.00
N sp2	=NO	6.67
N aromatic	:C2	14.01
N aromatic	:C:N	4.98
N(+) sp3	H3C	2.77
N(+) sp3	H2C2	82.36
N(+) sp2	C=CO(−)	68.61
N(+) sp2	C=NO	26.37
N(+) sp2	C=NO(−)	11.30
N(+) sp2	CO=O(−)	4.38
N(+) sp2	=CO2(−)	2.17
N(+) sp2	NO=O(−)	0.15
N(+) sp2	O2=O(−)	6.00
N(+) aromatic	H:C2	46.79
N(+) aromatic	:C2O(−)	7.10
N(+) sp	C#C(−)	14.36
N(+) sp	#CO(−)	0.00
N(+) sp	=N2(−)	19.14
O	HC	4.49
O	HC(pi)	8.19
O	HN(pi)	2.28
O	HO	29.95
O	C2	39.23
O	C2(pi)	31.33
O	C2(2pi)	24.06
O	CN(pi)	0.00
O	CN(+)(pi)	0.00
O	CN(2pi)	4.91
O	CO(pi)	27.16
O	CP(pi)	16.12
O	N2(2pi)	5.87
O	N2(+)(2pi)	6.27
P3	C3	16.70
P3	S3	66.68
P4	C3=N	0.00
P4	C3=O	30.50
P4	C3=S	46.30
P4	O3=O	0.00
S2	HC	2.58

Table 5.2 (continued)

Atom Type	Neighbors	Contribution
S2	HC(pi)	18.47
S2	C2	22.69
S2	C2(pi)	15.86
S2	C2(2pi)	7.94
S2	CN(pi)	25.96
S2	CN(2pi)	6.82
S2	CS(pi)	6.16
S2	CP(pi)	0.00
S2	N2	2.00
S2	N2(2pi)	21.36
S2	NS	1.00
S4	C2=O	5.89
S4	C2=O2	4.26
S4	CN=O2	9.20
Si	C4	2.02
Si	C3Si	0.67
H	H Acceptor[a]	8.63
Alkane	No. of C atoms[b]	0.53
Unsaturated HC	No. of C atoms[c]	0.10

[a] Intramolecular H bridge between acidic H (on O, N or S) and basic acceptor (O, N or F).
[b] Correction factor per carbon atom in pure alkanes.
[c] Correction factor per carbon atom in pure aromatics, olefins and alkynes.

(c) According to Tables 5.1 and 5.2, 1,3-diamino-2,4,6-trinitrobenzene has one (Atomic type = C aromatic, Neighbours = H:C2, and Contribution = 5.36 kJ/mol), three (Atomic type = C aromatic, Neighbours = :C2N(+), and Contribution = 28.67 kJ/mol), two (Atomic type = C aromatic, Neighbours = C:C:N, and Contribution = 11.10 kJ/mol), and three (Atomic type = N(+) sp2, Neighbours = CO=O(−), and Contribution = −4.38 kJ/mol).

$$\Delta_{sub}H = 21.03 + \sum_i a_i A_i + \sum_j g_j G_j$$

$$= 21.03 + 1(5.36) + 3(28.67) + 2(11.10) + 3(-4.38) = 121.46 \text{ kJ/mol}$$

The measured $\Delta_{sub}H$ for 1,3,5-trinitrobenzene, hexanitroethane, and 1,3-diamino-2,4,6-trinitrobenzene is 107.3, 70.7, and 143.5 JK^{-1} mol^{-1}, respectively [142].

5.2 Quantum mechanical and complex approaches for predicting the heat of sublimation

Several quantum mechanical calculations have been introduced to predict the heats of sublimation of energetic compounds [116, 119, 121, 275–277], and with which Politzer and coworkers have achieved significant success [121]. They used three quantum mechanical parameters in their calculations:

(1) the surface area of the 0.001 electron/bohr^3 isosurface of the electron density of the molecule,
(2) a measure of the variability of electronic potential on the surface,
(3) the degree of balance between the positive and negative charges on the isosurface.

Rice et al. [119] further improved this method to generate surface electrostatic potentials of individual molecules. Byrd and Rice [116] have modified previous methods by incorporating group additivity and by the use of the more complicated 6-311++G(2df,2p) basis set. Hu and coworkers [375] have also used the empirical relations of Politzer et al. [121] to predict the heats of sublimation of the condensed phases of energetic materials. There are also some relationships between the heats of sublimation of some polynitro compounds and lattice energies [278]. Suntsova and Dorofeeva [96] improved the electrostatic potential model of the Politzer approach by additional parameter Π (average deviation of electrostatic potential) for estimating enthalpies of sublimation of nitrogen-rich energetic compounds based on experimental enthalpies of sublimation for 185 compounds. Meftahi et al. [95] have compared several QSPR methods based on complex descriptors for predicting the enthalpy of sublimation of organic compounds containing energetic materials.

5.3 The use of structural parameters

There are several simple methods for predicting $\Delta_{sub}H$ which are based on structural features [89, 98–100]. These methods can be applied for selected classes of energetic compounds and are demonstrated here.

5.3.1 Nitroaromatic compounds

For nitroaromatics, the following optimized correlation can be used to predict $\Delta_{sub}H$ according to [98]:

$$\Delta_{sub}H = 64.51 + 4.555a - 2.763b + 10.32c + 16.51C_{SG}, \tag{5.3}$$

where $\Delta_{sub}H$ is in kJ/mol and the variable C_{SG} shows the contribution of certain polar groups. The different values of C_{SG} for various polar groups attached to nitroaromatic rings are specified as follows:

(1) Alkoxy group (–OR) attached to a nitroaromatic ring: $C_{SG} = 1.0$, e. g. 2-methoxy-1,3,5-trinitrobenzene.
(2) Carbonyl in form of –C(=O)NRR′ or –C(=O)–R attached to an aromatic ring in which R and R′ are alkyl groups: $C_{SG} = 0.75$, e. g. 3-nitroacetophenone.

(3) Carboxylic acid functional group, two hydroxyl groups, or three amino groups: $C_{SG} = 2.0$. In the case of two hydroxyl groups, nitro groups should be separated from −OH by at least one −CH− group, e. g. 4-nitrobenzene-1,2-diol.

Example 5.2. The use of equation (5.1) for the following molecular structure gives

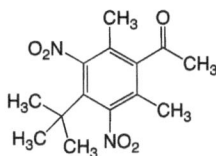

$$\Delta_{sub}H = 64.51 + 4.555a - 2.763b + 10.32c + 16.51C_{SG}$$
$$= 64.51 + 4.555(14) - 2.763(18) + 10.32(2) + 16.51(0.75)$$
$$= 111.6 \, kJ/mol.$$

The measured $\Delta_{sub}H$ value for this compound is 107.9 kJ/mol [279].

5.3.2 Nitramines

It was found that the molecular weight and structural parameters are sufficient to establish a new correlation as follows [98]:

$$\Delta_{sub}H = 15.62 + 0.3911Mw + 10.36n_{O_2NNCH_2NNO_2}, \tag{5.4}$$

where Mw is the molecular weight of the nitramine and $n_{O_2NNCH_2NNO_2}$ is the number of −CH$_2$− groups between two nitramine functional groups in cyclic and noncyclic nitamines. Equation (5.4) cannot be used for cyclic nitramines with $n_{O_2NNCH_2NNO_2} \geq 5$.

Example 5.3. 4,10-Dinitro-2,6,8,12-tetraoxa-4,10-diazaisowurtzitane (TEX) has the following molecular structure:

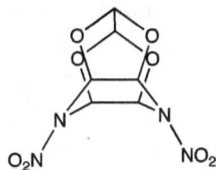

Equation (5.4) predicts $\Delta_{sub}H$ to be

$$\Delta_{sub}H = 15.62 + 0.3911Mw + 10.36n_{O_2NNCH_2NNO_2}$$
$$= 15.62 + 0.3911(262.13) + 10.36(0)$$
$$= 118.1 \, kJ/mol.$$

The measured $\Delta_{sub}H$ for TEX is 123.4 kJ/mol [238].

5.3.3 Nitroaromatics, nitramines, nitroaliphatics and nitrate esters

For nitroaromatics, nitramines, nitroaliphatics and nitrate esters, the molecular weight and the contribution of some specific functional groups, as well as structural features can be combined by a general equation as follows [99]:

$$\Delta_{\text{sub}}H = 53.74 + 0.2666Mw' + 13.99C_{\text{In}} - 15.58C_{\text{De}}, \tag{5.5}$$

where Mw' is the molecular weight of the nitro compound (except halogenated nitroaromatics and hydrogen-free nitro compounds in which the contribution of halogen atoms in the calculation of the molecular weight should be neglected), C_{In} is the contribution of specific polar groups attached to aromatic rings and C_{De} shows the presence of some molecular moieties. The values of C_{In} and C_{De} are specified according to the following conditions.

(1) Nitroaromatics:

    (a) Prediction of C_{In}:

        (i) –COOH and –OH functional groups: C_{In} = 2.0 for the compounds that contain the carboxylic acid functional group or two hydroxyl groups. Since the participation of a group in intramolecular hydrogen bonding can reduce its ability to form intermolecular hydrogen bonds, the presence of a nitro group in the *ortho* position relative to the –OH group can cancel its effect. If two hydroxyl groups are attached to the aromatic ring, nitro groups should be separated from –OH groups by at least one –CH– group, e. g. 4-nitrobenzene-1,2-diol.

        (ii) Amino groups: the value of C_{In} is equal to the number of amino groups in such compounds, i. e. $C_{\text{In}} = n_{\text{NH}_2}$.

        (iii) The presence of a carbonyl group in the form of an amide $\overset{\overset{\displaystyle O}{\|}}{\underset{\text{Ar} \quad \text{N}}{}}$ or ketone $\overset{\overset{\displaystyle O}{\|}}{\underset{\text{Ar} \quad \text{R}}{}}$: C_{In} = 0.75. Since carbonyl groups are in resonance with the aromatic ring, they can likely promote co-planarity and rigidity in some cases.

    (b) Prediction of C_{De}: The presence of alkyl groups – especially bulky groups such as tert-butyl – can decrease the intermolecular interactions for high ratios of $n_{\text{R}}/n_{\text{NO}_2}$. For $n_{\text{R}}/n_{\text{NO}_2} \geq 1$, the contribution of C_{De} should be considered. The values of C_{De} are 2.0 and 3.0 for the presence of one and more than one bulky group, respectively. If only small alkyl groups such as methyl groups are present, C_{De} = 1.0.

(2) Nitramines: The contributions of C_{In} and C_{De} depend on the number of N–NO$_2$ groups in cyclic and acyclic nitramines. For five membered (or larger) cyclic nitramines that have only the fragments $\underset{\text{H}_2\text{C}-\text{N}-\text{CH}_2}{\overset{\text{NO}_2}{|}}$ and for acyclic nitramines, $C = 1.75n_{\text{NNO}_2} - 4$. If $C < 0$ and $C > 0$, then C will become C_{De} and C_{In}, respectively.

For nitramines with the molecular fragment ![HN-C(=O)-NH fragment], appreciable molecular interactions are present so that $C_{In} = 4.25$.

(3) Nitroaliphatic compounds: For nitroaliphatic compounds, $C_{De} = 3.0$.

Example 5.4. The use of equation (5.5) for the following molecular structure gives

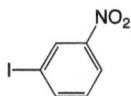

![nitrobenzene structure with NO2 and I substituents]

$$\Delta_{sub}H = 53.74 + 0.2666Mw + 13.99C_{In} - 15.58C_{De}$$
$$= 53.74 + 0.2666(122.10) + 13.99(0) - 15.58(0)$$
$$= 86.29 \text{ kJ/mol.}$$

The measured value of $\Delta_{sub}H$ is 83.0 kJ/mol [142].

5.3.4 General method for polynitro arenes, polynitro heteroarenes, acyclic and cyclic nitramines, nitrate esters, nitroaliphatics, cyclic and acyclic peroxides, as well as nitrogen-rich compounds

The following correlation can be used to predict $\Delta_{sub}H$ for a wide range of energetic compounds, including polynitro arenes, polynitro heteroarenes, acyclic and cyclic nitramines, nitrate esters, nitroaliphatics, cyclic and acyclic peroxides, as well as nitrogen-rich compounds [100]:

$$\Delta_{sub}H = 52.89 + 0.2689Mw' + 15.13F_{attract} - 13.29F_{repul}, \tag{5.6}$$

where $F_{attract}$ and F_{repul} are two parameters which take into account attractive and repulsive intermolecular forces. The values of $F_{attract}$ and F_{repul} are specified according to the following conditions, depending on the presence of various functional groups and molecular moieties:

(1) –COOH, –OH, ![N-O ring fragment], –NH$_2$ (or –NH–) and ![HN-C(=O)-NH] polar groups: The values of $F_{attract}$ are as follows:

(a) $F_{attract} = 2.0$ for compounds containing at least one –C(=O)OH, or two – OH functional groups. For nitroaromatics containing two –OH groups, nitro groups should be separated from –OH at least by one =CH– group, e. g. 4-nitrobenzene-1,2-diol. The presence of a nitro group in the *ortho* position with respect to the –OH group can cancel this condition so that $F_{attract} = 0.0$ in these compounds, e. g. 2,4,6-trinitro-1,3-benzenediol (styphnic acid).

(b) The value of $F_{attract}$ is equal to the number of –NH$_2$, –NH– or ![N-O ring fragment] groups.

(c) For nitramines or organic polynitrogen compounds which contain a $HN\diagdown\diagup NH$ molecular fragment, there is a large intermolecular attraction so that $F_{attract}$ = 4.7.

(2) Nitramines:

(a) Direct electrostatic interactions are dominant in polynitramine crystals which contain the $NO_2\ NO_2$ molecular fragment. Thus, $F_{attract}$ is 1.0 and 2.0 in acyclic and cyclic nitramines, respectively.

(b) For other acyclic nitramines, F_{repul} = 2.6.

(3) Nitroaliphatic compounds: The value of F_{repul} is equal to 2.6 in these compounds.

(4) Alkylated nitroaromatics: F_{repul} = 1.0 in substituted nitroaromatics containing small alkyl groups such as methyl groups. However, F_{repul} = 2.0 for bulky groups such as the t-butyl group. These conditions can be applied if $n_R/n_{NO_2} \geq 1$.

(5) Peroxides: For acyclic and cyclic peroxides, F_{repul} = 2.0.

(6) Intramolecular hydrogen bonding (H-bonding): F_{repul} = 0.5, e. g. 1,3-diamino-2,4,6-trinitrobenzene (DATB).

The conditions listed above are summarized in Table 5.3.

Table 5.3: Summary of correcting functions $F_{attract}$ and F_{repul}.

Molecular moieties	Effect on predicted $\Delta_{sub}H$		Comment
	$F_{attract}$	F_{repul}	
−OH and −COOH groups	2.0	−	(a) For one or more −COOH groups (b) For two −OH groups
−NH$_2$, −NH− and $N\diagdown N^+\text{−}O^-$ groups	No. of groups	−	−
HN$\diagdown\diagup$NH structure	4.7	−	−
NO$_2$ NO$_2$ structure	1.0	−	In acyclic nitramines
	2.0	−	In cyclic nitramines
−NO$_2$ and >NNO$_2$ groups	−	2.6	In nitroaliphatics and acyclic nitramines which are not included in the conditions two rows above
R$\diagup$−NO$_2$ structure, $n_R/n_{NO_2} \geq 1$	−	1.0	For small alkyl groups
	−	2.0	For bulky alkyl groups
C−O−O−C group	−	2.0	−
Intermolecular H-bonding	−	0.5	Intermolecular hydrogen bonding forms a 6-membered ring

Example 5.5. 1,3,5-Trinitro-1,3,5-triazinane (RDX) has the following molecular structure:

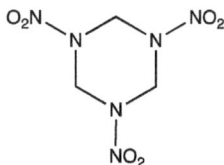

The use of equation (5.6) for the following molecular structure gives

$$\Delta_{sub}H = 52.89 + 0.2689 M_{rev} + 15.13 Mw' - 13.29 F_{repul}$$
$$= 52.89 + 0.2689(222.12) + 15.13(2.0) - 13.29(0)$$
$$= 142.9 \text{ kJ/mol.}$$

The deviation of the predicted value from the experimentally determined value (134.3 kJ/mol [280]), i. e. 8.3 kJ/mol, is lower than that of the values obtained from the two complex quantum mechanical calculations of Rice et al. [119] ($\Delta_{sub}H$ = 102.5 kJ/mol; Dev = 31.8) and [116] ($\Delta_{sub}H$ = 97.9 kJ/mol; Dev = 36.4).

5.4 Summary

This chapter has introduced different empirical methods for the prediction of the heats of sublimation of important classes of energetic compounds. Equation (5.2) can provide a group additivity method for prediction of the enthalpy of sublimation of organic compounds containing energetic materials. Equations (5.1) and (5.4) provide two simple and reliable approaches to estimate the heat of sublimation of nitoaromatics and nitramines, respectively. Equations (5.5) and (5.6) provide more complex empirical methods which can be applied to a wide range of organic compounds containing important energetic functional groups. Equation (5.6) is the best method because it can be used for a wide range of energetic compounds, including polynitro arenes, polynitro heteroarenes, acyclic and cyclic nitramines, nitrate esters, nitroaliphatics, cyclic and acyclic peroxides, as well as nitrogen-rich compounds.

6 Impact sensitivity

An organic energetic compound is metastable molecule, which is capable of undergoing very rapid and highly exothermic reactions. Thus, prediction of its sensitivity is a complex matter. Several properties contribute to a materials response to the stimulus in a sensitivity test, and which are a consequence of the kinetics and thermodynamics of the thermal decomposition of the explosive. They include

(1) the ease with which a detectable reaction of any kind can be initiated in an energetic compound;
(2) the tendency a small reaction can grow to destructive properties;
(3) the ease with which a higher order detonation can also be established in an energetic compound.

There are several reviews in literature which describe the calculation of the sensitivity of energetic compounds [81, 281–286]. Zeman and Jungová [285] have given an overview of the main developments in the study of the sensitivity of energetic materials to impact, shock, friction, electric spark, laser beams and heat in the period 2006–2015.

The safe handling of an energetic compound is one of the most important issues to the scientists and engineers who handle energetic molecules. Some stimuli can cause detonation of an energetic compound, including impact, shock, heat, electrostatic charge and friction. Of these stimuli, impact is probably the most well-known out of the many kinds of sensitivity because the drop-weight impact test is extremely easy to implement. Impact is one of the important factors in assessing an energetic compound since it provides information on the vulnerability of an energetic material to detonation due to accidental impact. Therefore, the impact sensitivity is closely related and highly relevant to many accidents in work places.

The drop hammer is one of the usual tests which is used for the evaluation of impact sensitivity. In this test, milligram quantities of an explosive material are placed between the flat tool steel anvil and flat surface of the tool striker. It typically involves

https://doi.org/10.1515/9783110740158-006

dropping a 2.5 kg mass from a predetermined height onto the striker plate. The impact drop height (H_{50}, cm) is the height from which there is a 50 % probability of causing an explosion, where 1 cm = 0.245 J (Nm) with a 2.5 kg dropping mass. Since the sensitivity is inversely proportional to H_{50}, impact sensitivity is shown in terms of the value of H_{50}. Although the impact sensitivity test is itself is extremely easy to implement, obtaining reliable experimental data is known to be relatively difficult. Since there is some difficulty associated with the initiation mechanism of explosion caused by mechanical impact, it can be assumed that hot spots in the material contribute to initiation in the drop weight impact test. The results of impact sensitivity are often not reproducible because factors in the test that might affect the formation and growth of hot spots can strongly affect the measurements. Moreover, the experimental data are extremely sensitive to the conditions under which the tests are performed. Despite all of the uncertainties associated with the impact sensitivity test, there are many different methods have been developed to correlate the impact sensitivity with other properties of energetic compounds, e. g. maximum heat of detonation [287], crystal lattice compressibility/free space [288], the available free space per molecule in the unit cell [289], ^{15}N NMR chemical shifts [290], nucleus-independent chemical shifts for aromatic explosives [291] and activation energy of thermal decomposition [292]. In recent years, some new correlations have been introduced to predict the impact sensitivity of different categories of energetic compounds, which are based on structural moieties [293–295], quantitative structure–property relationships (QSPR) [296–301], artificial neural networks and genetic algorithms [302, 303]. Some of these approaches are reviewed and illustrated in this chapter.

6.1 Complex methods

High speed computers allow quantum mechanical calculations of impact sensitivities of different classes of energetic compounds. Since the molecular surface of electrostatic potentials of the nitroaromatic molecules have positively charged regions over the $C-NO_2$ bonds, some authors have used computed partial atomic charges [304, 305], heats of reaction [306] and heats of explosion [307] in order to estimate impact sensitivities of some classes of explosives.

Brinck et al. [308] introduced the term "polarity index" (Π), which can measure local polarity, and demonstrated its relationship to the dielectric constant. For nitroaromatics, there is a relationship between Π and their impact sensitivities. Xiao and coworkers [309] proposed the thermodynamic criteria of "the smallest bond order", "the principle of the easiest transition" and the kinetic criterion of "the reaction activation energy of pyrolysis initiation" to judge the impact sensitivity. These methods are only used to qualitatively compare the relative magnitudes of impact sensitivity.

Politzer et al. [310] have identified a few features of electrostatic potentials for $C_aH_bN_cO_d$ explosives that appear to be related to their sensitivity to impact. Owen et al. [304] investigated the electrostatic potential over the C–NO$_2$ bonding region, which reflects a degree of instability in the C–NO$_2$ bond. For 18 nitroaromatics (excluding hydroxynitroaromatic molecules), Murray et al. [305] also introduced a correlation between impact sensitivity measurements and an approximation of the electrostatic potential at the midpoint of the C–N bond. Rice and Hare [311] used approximations of the electrostatic potential at midpoints, statistical parameters of these surface potentials and the property-structure relation method "generalized interaction property function" (GIPF) or computed heats of detonation to predict the impact sensitivity of $C_aH_bN_cO_d$ explosives. The impact sensitivities of $C_aH_bN_cO_d$ explosives have some dependence on the degree of internal charge imbalance within the molecule [312]. For nitramines, rupture of N–NO$_2$ bond is a key step in the decomposition process initiated by heat, shock and impact [313]. Edwards et al. [314] also used model IV of Rice and Hare [311] to calculate the heat of detonation of several nitramines using quantum mechanical theory, in which it was shown that there was a correlation between the exponential decrease of the HOMO and LUMO energies versus sensitivity at the DFT level of theory. Ren et al. [315] used seven models that related the features of molecular surface electrostatic potentials above the bond midpoints and rings, statistical parameters of surface electrostatic potentials to the experimental impact sensitivities of eight strained cyclic explosives with the C–NO$_2$ bonds at the DFT-B3LYP/6-311++G** level. Oliveira and Borges Jr [316] developed four mathematical models to correlate impact sensitivity to molecular charge properties using DFT.

Zhang et al. [317] derived some relationships between impact sensitivity and nitro group charges. They used the general gradient approximation (GGA) as well as the Beck hybrid functional and DNP basis set to calculate the Mulliken charges, which could be correlated with the impact sensitivity of nitro compounds. It was found that nitro compounds may be sensitive ($H_{50} \leq 40$ cm) when the nitro group has a negative charge of less than about 0.23. Since the charges on the nitro group can be used to estimate the bond strength, oxygen balance and molecular electrostatic potential, compounds with higher Mulliken net charges at the nitro groups will be insensitive and show large H_{50} values. The method of Zhang et al. [317] can be applied to nitro compounds when the C–NO$_2$, N–NO$_2$ or O–NO$_2$ bond is the weakest in the molecule. Zeman and Jungová [286] have also reviewed some futher publications which used quantum mechanical approaches. Bondarchuk [318] developed a theoretical approach for the prediction of impact sensitivity of explosives based on the solid state derived criteria, which include triggering pressure, the average number of electrons per atom, crystal morphology, energy content, and melting temperature. Cawkwell and Manner [319] demonstrated that chemical reactivity rather than thermomechanical effects is the dominant factor on explosive behavior in an impact test. They suggested that quantum-based molecular dynamics simulations may be a reliable computational tool for screening explosives for drop-height impact sensitivity. Mathieu [320] correlated linearly impact sensitivity for several high explosives with their (detonation velocity)$^{-4}$ or equivalently with (detonation pressure)$^{-2}$ or

(Gurney energy)$^{-1}$, which originated from the primary role of the amount of chemical energy evolved per atom for both performance and sensitivity.

Neural network architectures have been recently used as prediction methodology for impact sensitivity. Cho and coworkers [321] utilized 17 molecular descriptors, which were composed of compositional and topological descriptors in an input layer and two hidden neurons in a hidden layer. Some structural parameters have also been used to predict impact sensitivity using an artificial neural network model by choosing only 10 molecular descriptors [322]. The final neural structure consists of the three layers input, output, and hidden. The network is composed of: 10 input nodes, fifteen hidden-layer neurons and a single output neuron corresponding to the impact sensitivity of explosive. The ten structural descriptors include (1) a/MW; (2) b/MW; (3) c/MW; (4) d/MW; indicator variables for (5) aromaticity; (6, 7) heteroaromaticity (N and O); (8) N–NO$_2$; (9) α-hydrogen; (10) salt. The connection weights of the network were adjusted iteratively using back propagation algorithm. The predictive ability of the artificial neural network was checked with 275 experimental data. Impact sensitivities of 14 explosives in the test set were also compared with five quantum mechaical models of Rice and Hare [311]. It was shown that this model can provide better predictions compared to the quantum mechanical models of Rice and Hare [311].

Wang et al. [323] used the QSPR model by combining both the electronic and topological characteristics (ETSI approach) of the molecules under analysis. Since they used mixed data from very different structures, their predictions are, at best, just an indication. Xu et al. [301] performed a QSPR study for the entire set of 156 structurally different energetic compounds to estimate the impact sensitivity of new energetic materials. These QSPR approaches however, do not allow an evaluation of the chemical physics of initiation.

6.2 Simple methods on the basis of molecular structure for neutral energetic compounds

In contrast to complex methods, simple empirical correlations have the advantages that neither complex quantum chemistry software, nor high speed computers need to be available for tedious computation. There are some simple relationships that relate impact sensitivities with measured and predicted molecular properties, e. g. the oxygen balance of the molecules [324, 325], molecular electronegativities [326, 327], vibrational states [328], parameters related to oxidation numbers [329], ^{15}N NMR chemical shifts and heat of fusion [238, 281, 330], as well as elemental composition and molecular structures [293, 295, 331–335]. Several simple correlations, which can be applied to different classes of energetic compounds, are reviewed in the following sections.

6.2.1 Oxygen balance correlations

Kamlet and Adolph [324, 325] found reasonable linear correlations between the oxygen balance and $\log H_{50}$ for some classes of high energy molecules with similar decomposition mechanisms:

(1) Nitroaromatic:

$$\log H_{50} = 1.73 - 0.32 OB_{100} \tag{6.1}$$

(2) Nitroaromatic with α-CH linkage (e. g. TNT):

$$\log H_{50} = 1.33 - 0.26 OB_{100} \tag{6.2}$$

(3) Nitroaliphatic:

$$\log H_{50} = 1.74 - 0.28 OB_{100} \tag{6.3}$$

(4) Nitramine:

$$\log H_{50} = 1.37 - 0.17 OB_{100} \tag{6.4}$$

where $OB_{100} = 100(2d' - b' - 2a' - 2n'_{COO})$ in which d', b', a' and n'_{COO} are the number of oxygen, hydrogen, carbon and carboxylate entities in the molecule divided by molecular weight of the explosive.

6.2.2 Elemental composition and molecular moieties

6.2.2.1 Polynitroaromatics (and benzofuroxans) and polynitroaromatics with α-CH and α-N–CH (e. g. tetryl) and nitramines

It was shown that the following simple equations are suitable for polynitroaromatics (and benzofuroxans) and polynitroaromatics with α-CH and α-N–CH (e. g. tetryl), as well as for nitramines [331]:

(1) Polynitroaromatics (and benzofuroxans):

$$\log H_{50} = 11.8a' + 61.72b' + 26.9c' + 11.5d'. \tag{6.5}$$

(2) Polynitroaromatics with α-CH and α-N–CH (e. g. tetryl) and nitramines:

$$\log H_{50} = 47.3a' + 23.5b' + 2.36c' - 1.11d'. \tag{6.6}$$

Example 6.1. Consider 2,3,4,5-Tetranitrotoluene (2,3,4,5-TetNT) with the following molecular structure:

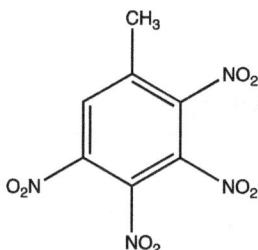

Since it is a nitroaromatic compound with a α-CH linkage, equation (6.6) can be used, which gives

$$\log H_{50} = 47.3a' + 23.5b' + 2.36c' - 1.11d'$$
$$= (47.3(7) + 23.5(4) + 2.36(4) - 1.11(8))/272.13$$
$$= 1.565$$
$$H_{50} = 37 \text{ cm}.$$

The measured H_{50} for this compound is 15 cm [311].

6.2.2.2 Nitroaliphatics, nitroaliphatics containing other functional groups and nitrate explosives

For nitroaliphatics, nitroaliphatics containing other functional groups and nitrate explosives, the following correlation can be used to predict their impact sensitivity [333]:

$$\log H_{50} = 2.5 + 0.371[100(a' + b'/2 - d')]$$
$$- 0.485(100c') + 0.185n_{R-C(NO_2)_2-CH_2-}, \qquad (6.7)$$

where $n_{R-C(NO_2)_2-CH_2-}$ is the number of $R-C(NO_2)_2-CH_2-$ groups attached to the oxygen atom of carboxylate functional groups (R is an alkyl group).

Example 6.2. The use of equation (6.7) for bis-(2,2-dinitropropyl) carbonate with the following molecular structure gives

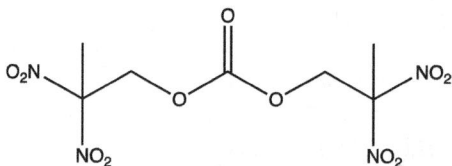

$$\log H_{50} = 2.5 + 0.371[100(a' + b'/2 - d')]$$
$$- 0.485(100c') + 0.185n_{R-C(NO_2)_2-CH_2-}$$
$$= 2.5 + 0.371[100(7 + 10/2 - 11)/326.17]$$

$$- 0.485(100 \times 4/326.17) + 0.185(2)$$

$$= 2.389$$

$$H_{50} = 228 \text{ cm.}$$

The measured H_{50} for this compound is 300 cm [336]. If equation (6.3) is used, there is a large deviation between the predicted value and the experimental value of 121 cm (Dev = 179 cm).

6.2.2.3 Nitroheterocycles

For nitroheterocyclic energetic compounds including nitropyridines, nitroimidazoles, nitropyrazoles, nitrofurazanes, nitrotriazoles and nitropyrimidines, the following general equation can be used for various types of $C_aH_bN_cO_d$ nitro heterocycles [332]:

$$\log H_{50} = 46.29a' + 35.63b' - 7.700c' + 7.943d'$$
$$+ 44.42n'_{-CNC-} + 102.3n'_{-CNNC-}, \tag{6.8}$$

where n'_{-CNC-} and n'_{-CNNC-} are the number of –CNC– and –CNNC– moieties in the aromatic ring divided by the molecular weight of the explosive.

Example 6.3. If equation (6.8) is used for 4-methyl-3,5-dinitro-1,2,4-triazole with the following molecular structure,

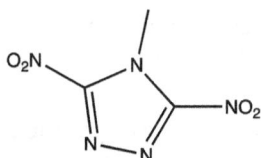

the value of H_{50} is calculated as follows:

$$\log H_{50} = 46.29a' + 35.63b' - 7.700c' + 7.943d'$$
$$+ 44.42n'_{-CNC-} + 102.3n'_{-CNNC-}$$
$$= (46.29(3) + 35.63(3) - 7.700(5) + 7.943(4)$$
$$+ 44.42(1) + 102.3(1))/173.09$$
$$= 2.229$$
$$H_{50} = 169 \text{ cm.}$$

The measured H_{50} for this compound is 155 cm [336]. It was found that the complex neural network [321] approach results in a larger deviation between the predicted value and experimental data, i. e. 64 cm (Dev = 91 cm).

6.2.2.4 Polynitroheteroarenes

An improved correlation with respect to equation (6.8) has been introduced to predict the impact sensitivity of different types of polynitroheteroarenes including nitropyridine, nitroimidazole, nitropyrazole, nitrofurazane, nitrooxadiazole, nitro-1,2,4-triazole, nitro-1,2,3-trazole and nitropyrimidine explosives as [334]:

$$\log H_{50} = 52.13a' + 31.80b' + \frac{117.6 \sum SSP_i}{Mw}, \tag{6.9}$$

where SSP_i are specific structural parameters that can decrease or increase impact sensitivity. The values of SSP_i are specified according to the molecular structures as follows:

(1) Amino derivatives as substituents in heteroarene: Amino derivatives (Ar–NH– or R–NH–) can decrease the impact sensitivity of some explosives [206]. For heteroarenes containing tetrazole derivatives or three consecutive nitrogen atoms (e. g. 1,2,3-triazole derivatives) or the nitropyrimidine group attached to an aromatic ring via nitrogen (e. g. 1-picryl-2-picrylamino-1,2-dihydropyrimidine), the presence of amino groups has no notable effect and the value of SSP_i is zero. For the presence of amino groups in nitropyridine, nitrofurazane (or nitrooxidazole) and nitro-1,2,4-triazole (or nitropyrimidine) explosives, the values of SSP_i are 0.5, 0.6 and 2.5 respectively. For other nitroheteroarenes which have only nitrogen as a heteroatom, SSP_i is 2.0 in the presence of amino derivatives.

(2) The attachment of an aromatic ring (e. g. picryl) to nitrogen and the presence of one nitro group in a specific position:

   (a) The values of SSP_i are equal to 1.0 and 0.6 for the attachment of an aromatic ring to nitrogen in imidazole (or only in mononitro imidazol) and nitropyrazole explosives, respectively.

   (b) If only two aromatic substituents are attached to the heteroarene ring without further substituents, $SSP_i = 2$.

   (c) If the polynitrophenyl group is attached to the nitrogen atom at the 4-position in 1,2,4-triazole explosives (e. g. 4-(2,4-dinitrobenzyl)-3,5-1,2,4-triazole), $SSP_i = 0.7$.

   (d) The values of SSP_i are 0.6, 0.8 and 1.0 for the presence of one nitro group in positions of 2 in 3 (or 5) in nitropyrazole, nitroimidazole and in position 3 in nitro-1,2,4-triazole explosives, respectively. This condition is valid for nitroimidazoles up to only disubstitueted nitroimidazole explosives.

   As an illustrative example for this section, $\sum SSP_i$ is equal to 1.8 for 2-nitro-1-picryl-imidazole, whereas the $\sum SSP_i$ value is 1.0 for the isomer 4-nitro-1-picryl-imidazole.

Less sensitive materials could be designed if the "trigger linkage" could be identified and avoided [337]. Ammonium salts are "unusually stable" because when an acid is converted to its ammonium salt. Thus, the ammonium salt will be less sensitive than the parent acid [336]. Since sensitivity of 1,2,4-triazoles are usually low, it appears that the ammonium counterion in ammonium 3,5-dinitro-1,2,4-triazolate provides no special insensitivity. For less sensitive derivatives of nitroimidazole, nitropyrazole, and nitro-1,2,4-triazole explosives, the insensitivity of the explosive is, in fact, a consequence of the chemistry proceeding the rate determining step. Due to considerable charge delocalization through $-N=N-$ and $-C=C-$ double bonds, the insensitivity to impact may be accounted for in some isomers of polynitroheteroarenes including nitroimidazole, nitropyrazole and nitro-1,2,4-triazole explosives.

(3) Nitro-1,2,3-triazole explosives: For the attachment of an aromatic ring to a nitrogen atom in position 1, the value of SSP_i is equal to -1.0. As was mentioned in part (1), this condition is valid for compounds which are not amino derivatives. The sensitivity to impact and instability varies from isomer to isomer in nitro-1,2,3-triazole explosives. Since there are large differences in the impact sensitivities of 1-picryl-1,2,3-triazole compared to 2-picryl-1,2,3-triazole, and of 4-nitro-1-picryl-1,2,3-triazole compared to 4-nitro-1-picryl-1,2,3-triazole, it is possible that this is due to the facile loss of nitrogen in the 1-picryl isomers [338].

For the presence of another more active site to initiate decomposition, e. g. $R-NO_2$, the value of $\sum SSP_i$ is taken as zero.

Example 6.4. 5,5'-Dinitro-1H,1'H-3,3'-bi(1,2,4-triazole) has the following molecular structure:

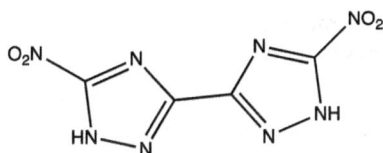

If equation (6.9) is used for this compound, it gives

$$\log H_{50} = 52.13a' + 31.80b' + \frac{117.6 \sum SSP_i}{Mw}$$

$$= (52.13(4) + 31.80(2))/226.11 + \frac{117.6(2)}{226.11}$$

$$= 2.243$$

$$H_{50} = 175 \, cm.$$

The measured H_{50} for this compound is 153 cm [336]. It was found that two complex neural networks of Cho et al. [321] and Keshavarz–Jafari [322] result in larger deviations between the predicted value and experimental data, i. e. 73 cm (Dev = 80 cm) and 200 cm (Dev = 47 cm), respectively. The use of equation (6.8) also results in a larger deviation, i. e. 200 cm (Dev = 47 cm).

6.2.2.5 Nitroaromatics, benzofuroxans, nitroaromatics with α-CH, nitramines, nitroaliphatics, nitroaliphatics containing other functional groups and nitrate energetic compounds

A simple correlation has been introduced to predict the impact sensitivity of nitroaromatics, benzofuroxans, nitroaromatics with α-CH, nitramines and nitroaliphatics, as well as nitroaliphatics containing other functional groups and nitrate explosives as [335]

$$\log H_{50} = 48.81a' + 25.94b' + 13.73c' - 4.786d'$$
$$+ \frac{111.6DSSPH - 132.3ISSPH}{Mw}, \tag{6.10}$$

where $DSSPH$ and $ISSPH$ are decreasing and increasing sensitivity structural parameters, respectively, which can be specified based on the molecular structures as follows.

(1) Prediction of $DSSPH$

    (a) Nitroaromatics and bezofuroxanes: Since the presence of some special electron donating substituents which have an electron pair located on the atom which attaches to the aromatic ring (such as $-NH_2$ and $-OCH_3$), or the presence of double and triple bonds involving the carbon atom which is attached to an aromatic ring (e. g. $-C(=O)-$ and $-CN$) can decrease the sensitivity, the effects of these substituents can be predicted based on the molecular structure.

        (i) $-NH_2$ group: $DSSPH$ equals 0.7, 1.2 and 1.7 for n_{NH_2} = 1, 2 and 3 per aromatic ring, respectively, if $n_{NO_2} \leq 3$ per aromatic ring.

        (ii) Only $-OH$ groups: If $n_{NO_2} = n_{OH}$, then $DSSPH$ = 0.9.

        (iii) Benzofuroxans: $DSSPH$ = 0.6.

        (iv) $-OR$ and $-O^-$ groups attached to aromatic ring: $DSSP$ is 0.7 and 0.5, respectively. If an $-NH_2$ group is also attached to the same aromatic ring, the value of $DSSPH$ is 1.2.

        (v) Double and triple bonds involving the carbon atom attached to the aromatic ring (e. g. $-C(=O)-$ and $-CN$): If $n_{NO_2} \geq 3$ in the aromatic ring, then $DSSPH$ = 1. If the 2,2-dinitropropyl group is attached to the $-COO-$ functional group, $DSSPH$ = 2.

        (vi) Nitroaromatic explosives that contain one methyl group: For the presence of one phenyl or one $-OH$ group in the meta position with respect to the methyl group, $DSSPH$ = 0.8.

    (b) Nitramines: For nitramines which contain the $=C-N-NO_2$ group, $DSSPH$ = 0.45.

    (c) Nitroaliphatics: For nitroaliphatics containing the $-COO-$ functional group, the number of 2,2-dinitropropyl and nitroisobutyl groups attached to $-COO-$ can increase the value of $DSSPH$ since the value of $DSSPH$ equals the number of 2,2-dinitropropyl and nitroisobutyl groups.

(2) Prediction of *ISSPH*
   (a) Nitroaromatics
      (i) If an α-CH linkage is attached to a nitroaromatic ring, *ISSPH* = 0.5. It should be mentioned that condition (1) (a) (vi) cannot be used here.
      (ii) For those nitroaromatics containing azido or diazo functional group, *ISSPH* = 0.7.
   (b) Nitramines: *ISSPH* = 0.5. The presence of the molecular structure given in condition (1) (b) has the reverse effect.

Example 6.5. Consider ammonium nitrate with the following molecular structure:

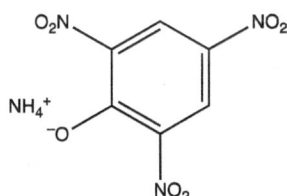

The use of equation (6.9) gives

$$
\begin{aligned}
\log H_{50} &= 48.81a' + 25.94b' + 13.73c' - 4.786d' \\
&\quad + \frac{111.6\,DSSPH - 132.3\,ISSPH}{Mw} \\
&= (48.81(6) + 25.94(6) + 13.73(4) - 4.786(7))/246.13 \\
&\quad + \frac{111.6(0.5) - 132.3(0)}{246.13} \\
&= 2.136
\end{aligned}
$$

$$H_{50} = 137\ \text{cm}.$$

The measured H_{50} for this explosive is 135 cm [336].

6.2.2.6 An improved simple model for the prediction of the impact sensitivity of different classes of energetic compounds

An improved simple model with respect to equation (6.10) has been introduced to predict the impact sensitivity of nitropyridines, nitroimidazoles, nitropyrazoles, nitrofurazanes, nitrotriazoles, nitropyrimidines, polynitro arenes, benzofuroxans, polynitro arenes with α-CH, nitramines, nitroaliphatics, nitroaliphatics containing other functional groups and nitrate energetic compounds as [293]:

$$(\log H_{50})_{\text{core}} = -0.584 + 61.62a' + 21.53b' + 27.96c' \tag{6.11}$$

$$\log H_{50} = (\log H_{50})_{\text{core}} + 84.47\frac{F^+}{Mw} - 147.1\frac{F^-}{Mw}. \tag{6.12}$$

The parameters F^+ and F^- in equation (6.12) are two correcting functions, which are described in the following sections.

Molecular fragments affecting F^+

(1) $-NH_2$ groups and amino derivatives as substituents:
   (a) Polynitro hetroarenes: The values of F^+ are 0.4, 0.6 and 3.0 for the presence of amino derivatives (Ar–NH– or R–NH–) in nitropyridine, nitrofurazane (or nitrooxidazole) and nitro-1,2,4-triazole (or nitropyrimidine) explosives, respectively.

   (i) For amino derivatives of other nitroheteroarenes which contain only nitrogen as heteroatoms, $F^+ = 2.3$.

   (ii) For heteroarenes which contain four nitrogen atoms (e. g. 1,2,3,4-tetrazole derivatives) or three nitrogens (e. g. 1,2,3-triazole derivatives) attached consecutively in one ring, or nitropyrimidine explosives in which an aromatic ring is attached to a nitrogen atom (e. g. 1-picryl-2-picrylamino-1,2-dihydropyrimidine), $F^+ = 0.0$

   (b) Polynitro arenes: For $n_{NO_2} \leq 3$ per aromatic ring, $F^+ = 0.7$, 1.6 and 2.2 for $n_{NH_2} = 1$, 2 and 3 per aromatic ring, respectively.

(2) Molecular fragments that increase insensitivity:
   (a) Polynitro hetroarenes: The attachment of an aromatic ring (e. g. picryl) to nitrogen, or if one nitro group is present in a specific position.

   (i) The value of F^+ is equal to 1.1 and 0.6 for the attachment of an aromatic ring to nitrogen in imidazole (or only in mononitro imidazole) and nitropyrazole explosives, respectively.

   (ii) If only two aromatic substituents are present and attached to a heteroarene ring without further substituents, $F^+ = 2.8$.

   (iii) If the polynitrophenyl group is attached to the nitrogen atom at the 4-position in 1,2,4-triazole explosives (e. g. 4-(2,4-dinitrobenzyl)-3,5-1,2,4-triazole), $F^+ = 0.7$.

   (iv) The values of F^+ are 0.6, 0.9 and 1.0 if one nitro group is present in position 2 of nitroimidazoles, 3 (or 5) in nitropyrazoles and 3 in nitro-1,2,4-triazole explosives, respectively. This situation is valid for nitroimidazoles up to only disubstituted nitroimidazole explosives. For the presence of one carbonyl group, or the attachment of two 5-nitro-1,2,4-triazole rings in nitro-1,2,4-triazole explosives, the value of F^+ is equal to 1.4.

   (b) Polynitro arenes:

   (i) If only –OH groups are attached and if $n_{NO_2} = n_{OH}$, then $F^+ = 1.25$.

   (ii) If –OR and –O$^-$ groups are attached to the aromatic ring, $F^+ = 0.7$ and 0.5, respectively. If an $-NH_2$ group is also attached to the aromatic ring, $F^+ = 1.2$.

(iii) If the carbon attached to the aromatic ring participates in double or triple bonds (e. g. $-C(=O)-$ and $-CN$), and $n_{NO_2} \geq 3$ for the aromatic ring, then $F^+ = 1.0$. The value of F^+ equals 2.0 for the attachment of the 2,2-dinitropropyl group directly to $-COO-$ functional group.

(iv) For the attachment of one methyl group, if there is one phenyl or one OH group in the meta position with respect to the methyl group, $F^+ = 0.8$.

(c) Benzofuroxanes: $F^+ = 0.6$.

(d) Nitramines: For nitramines containing the group $=C-N-NO_2$, $F^+ = 0.7$.

(e) Nitroaliphatics: For nitroaliphatics containing the $-COO-$ functional group, the value of F^+ is equal to the number of 2,2-dinitropropyl and nitroisobutyl groups attached to the $-COO-$ group multiplied by 1.4. For 2,2-dinitropropanediol, $F^+ = 2.8$.

(3) Prediction of F^-:

(a) Nitro-1,2,3-triazole explosives: If an aromatic ring is attached to nitrogen in position 1, the value of F^- equals 1.0. This condition is valid for nonamino derivatives.

(b) Polynitro arenes:

(i) The presence of an α-CH linkage attached to carbocyclic nitroaromatic ring may increase the sensitivity, and therefore, $F^- = 0.5$. It should be mentioned that condition (2) (b) (iv) is an exception, that can decrease impact sensitivity.

(ii) For those polynitro arenes that contain azido or diazo functional group, $F^- = 0.7$.

(c) Nitramines: If the $N-NO_2$ functional group is present, $F^- = 0.5$. For nitramines that contain the $=C-N-NO_2$ group, condition (2) (d) does not apply.

The values of F^+ and F^- are summarized in Tables 6.1 and 6.2.

Example 6.6. For 2-picryl-1,2,3-triazole with the following molecular structure, the use of equations (6.11) and (6.12) gives

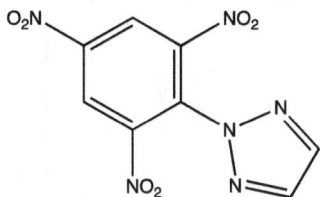

$$(\log H_{50})_{core} = -0.584 + 61.62a' + 21.53b' + 27.96c'$$
$$= -0.584 + (61.62(8) + 21.53(4) + 27.96(6))/280.15$$
$$= 2.082$$

Table 6.1: Prediction of F^+.

Molecular moieties	Compound	F^+	Illustration	Exception
Presence of –NH$_2$ groups and amino derivatives as substituents	Polynitro hetroarenes	0.4	Nitropyridine	For central heteroarenes which contain four nitrogens (e. g. 1,2,3,4-terazole derivatives) and three nitrogens attached consecutively in one ring (e. g. 1,2,3-triazole derivatives), or nitropyrimidine explosives in which an aromatic ring is attached to nitrogen (e. g. 1-picryl-2-picrylamino-1,2-dihydropyrimidine), the presence of amino derivatives has no notable effect and the value of F^+ can be taken as zero
		0.6	Nitrofurazane (or nitrooxidazole)	
		3.0	Nitro-1,2,4-triazole (or nitropyrimidine)	
		2.3	Other nitroheteroarenes which contain only nitrogen as hetero atoms	
	Polynitro arenes (if $n_{NO_2} \leq 3$ per each aromatic cycle)	0.7	n_{NH_2} = 1 per each aromatic cycle	
		1.6	n_{NH_2} = 2 per each aromatic cycle	
		2.2	n_{NH_2} = 3 per each aromatic cycle	
Molecular fragments which increase insensitivity	Nitroimidazole, nitropyrazole and nitro-	0.6	The attachment of aromatic ring to nitrogen in nitropyrazole	
	1,2,4-triazole explosives	1.1	The attachment of aromatic ring to nitrogen in imidazole (or in mononitro imidazole)	
		2.8	If only two aromatic substituents are attached to a heteroarene ring without further substituents	
	1,2,4-Triazole explosives	0.7	If polynitrophenyl is attached to the nitrogen atom at the 4-position (e. g. 4-(2,4-dinitrobenzyl)-3,5-1,2,4-triazole)	–
	Nitroimidazole	0.6	Presence of one nitro group in position 2	This condition is valid for nitroimidazole and disubstituted nitroimidazole explosives

Table 6.1 (continued)

Molecular moieties	Compound	F^+	Illustration	Exception
Molecular fragments which increase insensitivity	Nitropyrazole	0.9	One nitro group is present in position 3 (or 5)	
	Nitro-1,2,4-triazole	1.0	One nitro group is present in position 3	One carbonyl group is present, or the direct attachment of only two 5-nitro-1,2,4-triazole rings, the value of F^+ is equal to 1.4
	Polynitro arenes	1.25	Only one –OH group is attached: if $n_{NO_2} = n_{OH}$	–
		0.7	–OR is present	If –NH$_2$ group has been attached also to aromatic ring simultaneously, the value of F^+ becomes 1.2
		0.5	–O⁻ is present	
		1.0	The carbon atom which is attached to the aromatic ring participates in double or triple bonds (e. g. –C(=O)– and –CN), and if $n_{NO_2} \geq 3$ in aromatic ring	The value of F^+ is equal to 2.0 for the presence of 2,2-dinitropropyl attached to the –COO– functional group
		0.8	For the attachment of one methyl group, if there is one phenyl or one –OH group in the *meta* position with respect to methyl group	
	Benzofuroxanes	0.6		
	Nitroaliphatics		For nitroaliphatics that contain the –COO– functional group, the value of F^+ is equal to the number of 2,2-dinitropropyl and nitroisobutyl groups attached to –COO– times 1.4	If the 2,2-dinitropropanediol group is present, $F^+ = 2.8$

Table 6.2: The value of F^- for specific molecular groups.

Compound	F^-	Illustration	Exception
Nitro-1,2,3-triazole	1.0	The attachment of an aromatic ring to nitrogen in position 1	This condition is valid in the absence of amino groups
Polynitro arenes	0.5	The presence of an α-CH linkage attached to carbocyclic nitroaromatic ring	For the attachment of one methyl group, if there is one phenyl or one –OH group in the *meta* position with respect to the methyl group then $F^+ = 0.8$
	0.7	For those polynitro arenes that contain the azido or diazo functional group	
Nitramines	0.5		For nitramines that contain the group =C–N–NO$_2$, $F^+ = 0.7$

$$\log H_{50} = (\log H_{50})_{core} + 84.47\frac{F^+}{Mw} - 147.1\frac{F^-}{Mw}$$
$$= 2.082 + 84.47\frac{0}{280.15} - 147.1\frac{0}{280.15}$$
$$= 2.082$$
$$H_{50} = 121\ cm.$$

The experimental value of H_{50} for this compound is 200 cm [336]. It was found that the complex neural network method of Keshavarz–Jafari [322] give a value which showed a larger deviation between the predicted value and experimental data, i. e. 57 cm (Dev = 143 cm).

6.3 Impact sensitivity of quaternary ammonium-based energetic ionic liquids or salts

It was found that the impact sensitivity of quaternary ammonium-based energetic ionic liquids or salts can be correlated with the elemental composition of cations and anions and two correcting functions as follows [339]:

$$IS_{IL}(J)$$
$$= 35.04$$
$$+ \frac{-1073a_{cat} + 728.9b_{cat} - 1761d_{cat} + 1032a_{ani} + 1061b_{ani} - 1261c_{ani} - 944.0d_{ani}}{Mw}$$
$$+ \frac{3663IS_{IL}^+ - 4291IS_{IL}^-}{Mw} \tag{6.13}$$

where $IS_{IL}(J)$ is the impact sensitivity of a desired energetic of quaternary ammonium-based energetic ionic liquids or salts in J; a_{cat}, b_{cat} and d_{cat} are the number of carbon, hydrogen and oxygen atoms in the cation, respectively; a_{ani}, b_{ani}, c_{ani} and d_{ani} are the number of carbon, hydrogen, nitrogen and oxygen atoms in the anion, respectively; Mw is the molecular weight of the desired quaternary ammonium-based energetic ionic liquid or salt; IS_{IL}^{+} and IS_{IL}^{-} are two correction functions that depend on stabilizing or destabilizing structural parameters in cations or anions. The values of IS_{IL}^{+} and IS_{IL}^{-}, for the presence of specific cations or anions given in Table 6.3, are equal to 1.0.

Example 6.7. The use of equation (6.13) for the following compound

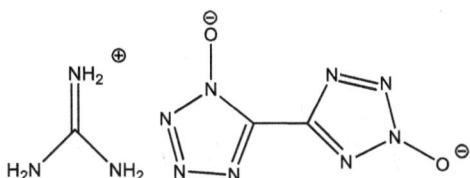

gives:

$$IS_{IL}(J)$$

$$= 35.04 + \frac{-1073 a_{cat} + 728.9 b_{cat} - 1761 d_{cat} + 1032 a_{ani} + 1061 b_{ani} - 1261 c_{ani}}{Mw}$$

$$+ \frac{-944.0 d_{ani} + 3663 IS_{IL}^{+} - 4291 IS_{IL}^{-}}{Mw}$$

$$= 35.04 + \frac{-1073 \times 1 \times 2 + 728.9 \times 2 \times 3 - 1761 \times 0 + 1032 \times 2 + 1061 \times 0 - 1261 \times 8}{228}$$

$$+ \frac{-944.0 \times 2 + 3663 \times 1 - 4291 \times 0}{228} = 36.26\,J$$

The measured value of IS_{IL} for this compound is 40 J [340].

6.4 Summary

This chapter demonstrates several simple empirical methods for the prediction of the impact sensitivity of important classes of energetic compounds. Equations (6.5), (6.6) and (6.7) are suitable correlations for predicting the impact sensitivity of
(1) Polynitroaromatics (and benzofuroxans),
(2) Polynitroaromatics with α-CH and α-N–CH linkages (e. g. tetryl) and nitramines,
(3) Nitroaliphatics, nitroaliphatics containing other functional groups and nitrate explosives.

Table 6.3: The contribution of IS_{IL}^+ and IS_{IL}^- in ionic liquids or salts.

Cation	Anion
IS_{IL}^+	
$R-\overset{R}{\underset{R}{\overset{\oplus}{N}}}-(CH_2)n-N_3$	$N(NO_2)^{2-}$
$R_1-\overset{R_2}{\underset{NH_2}{\overset{\oplus}{N}}}-NH_2$	$N(NO_2)(CN)^-$ or $C(NO_2)_2(CN)^-$ or NO_3^-
NH_4^+	(tetrazole/triazole structures) or
guanidinium ($C(NH_2)_3^+$)	(triazole/tetrazole structures) or
aminoguanidinium	(5-aminotetrazolate structure)
IS_{IL}^-	
($R_1R_2R_3N^+NH_2$) or (guanidinium type) or (triazenyl type)	(nitrotetrazolate) or (hydroxy tetrazole structure)
aminoguanidinium	ClO_4^-
NH_4^+	$N(NO_2)^{2-}$ or (aminotriazole/tetrazole structures)
$N_2H_5^+$	(nitro-triazole/tetrazole structures) or (structures) or

For nitroheterocyclic energetic compounds containing nitropyridines, nitroimida-zoles, nitropyrazoles, nitrofurazanes, nitrotriazoles or nitropyrimidines, equation (6.8) is a simple method for estimating the impact sensitivity of these classes of ni-tro heterocycles. Equation (6.9) is an improved correlation of equation (6.8), which can be used not only for the previously mentioned classes of polynitroheteroarenes, but also for two further important classes of energetic compounds, namely, nitro-1,2,4-triazole and nitro-1,2,3-trazole derivatives. Equation (6.10) is an extended corre-lation for the prediction of the impact sensitivity of nitroaromatics, benzofuroxans, nitroaromatics with α-CH linkages, nitramines and nitroaliphatics, as well as ni-troaliphatics containing other functional groups and nitrate explosives. Equations (6.11) and (6.12) introduce a general correlation that can be used for a wide range of the above-mentioned classes of energetic compounds, including: nitropyridines, nitroimidazoles, nitropyrazoles, nitrofurazanes, nitrotriazoles, nitropyrimidines, polynitro arenes, benzofuroxans, polynitro arenes with α-CH linkages, nitramines, nitroaliphatics, nitroaliphatic containing other functional groups and nitrate ener-getic compounds. Equation (6.13) provides a simple correlation for prediction of the impact sensitivity of quaternary ammonium-based energetic ionic liquids or salts.

7 Electric spark sensitivity

The electrostatic or electric spark sensitivity (E_{ES}) of an energetic compound is an important aspect for estimating its safety in an electrostatic discharge environment – and could be helpful in reducing accidents. It can be defined as the degree of sensitivity to an electrostatic discharge. It represents the ease with which an explosion can be initiated by an electrostatic spark.

7.1 Measurement of electric spark sensitivity

Electric spark sensitivity is determined by subjecting an energetic compound to a high-voltage discharge from a capacitor. The required energy is calculated from the known capacitance C (in F) of the circuit and voltage U (in V) at the condenser as

$$E_{ES} = 0.5CU^2. \tag{7.1}$$

Since the value of E_{ES} depends on the configuration of the electrodes and structure of the circuit, it can be expected that various results should be obtained by different test specifications of the electrode energy used by different authors [341]. Some of the important parameters for the determination of the E_{ES} include:
(1) the surface area of the tip of the electrode can affect the energy density of the spark;
(2) the structure of the electrical circuit will affect the shape and duration of the electrical pulse, which in turn influences the rate and duration of the delivery of energy to the sample;
(3) confinement of the sample (or lack thereof) can have a dramatic effect on the measured electrostatic sensitivity because of the tendency of some samples to form dust clouds when unconfined.

To determine the values of the electric spark sensitivity of some secondary explosives, Zeman and coworkers [341, 342] used an instrument marked as RDAD, which was constructed in the R&D Department of Zbrojovka Indet, Inc., Vsetín, Czech Republic

https://doi.org/10.1515/9783110740158-007

[341] for this purpose. Furthermore, they have measured the electric spark sensitivity for a large set of polynitro secondary explosives. The capacitance of the capacitor in the RDAD instrument is chosen so as to allow measurements in the voltage range of 8 to 14 kV. If the initiation of the explosive is successful, the next measurement is carried out at a voltage which is 0.2 kV lower. Whereas, if initiation was unsuccessful, the voltage is increased by the same value (up and down method). Fortunately, there is a linear relationship between the measured data obtained using the RDAD instrument with some of those obtained by other experimental methods used to determine the electric spark sensitivity in other recognized laboratories [341]. Since the RDAD instrument is not suitable for determining the sensitivity of primers and pyrotechnics [63], another instrument marked as ESZ KTTV has also been developed, with financial support from Czech Ministry of Industry and Commerce [342]. The ESZ KTTV instrument is suitable for both primary and secondary explosives. Details of the apparatus and procedures for the RDAD and ESZ KTTV instruments were described elsewhere [341–343].

7.2 Different methods for predicting electric spark sensitivity

For some classes of the secondary explosives, there are some correlations between the electric spark sensitivity and some characteristics, such as the detonation velocity and the Piloyan activation energy of decomposition which is obtained from differential thermal analysis (DTA) [344–348]. Zeman et al. [344, 348, 349] have indicated that there is a linear relationship between the electric spark sensitivity and the square of the detonation velocity for some specific categories of explosives. Zeman and Liu [350] have shown that the electric spark sensitivities of some nitramines are directly proportional to the crystal lattice-free volumes but there were several nitramines with the opposite course of this relationship. For 28 polynitroarenes and their derivatives, Zeman [351] demonstrated some relationships between their impact and electric spark sensitivities with the volume heats of their explosion and their enthalpies of formation. Tan et al. [352] correlated the electrostatic spark energy with four parameters – the standard deviation of the negative electrostatic potential, the minimum surface electrostatic potential, the minimum ionization energy, and the detonation pressure using genetic function approximation. Wang et al. [353, 354] used quantum chemistry methods to optimize the molecular geometries and electronic structures for some explosives. They have shown that there are quantitative relationships between the experimental electrostatic spark sensitivity values and the predicted detonation velocity and pressure for some special groups of explosives. Some new correlations have also been developed which use the maximum obtainable detonation pressure (or velocity) as well as some specific molecular fragments to predict the electric spark sensitivity [355–358]. These methods have the advantage that there is no need to use the crystal density and heat of formation of an explosives to predict its detonation pressure and velocity. A simple method was also introduced to correlate the electric spark and impact sensitivities of nitroaromatic compounds [359].

Türker [360] has used quantum chemistry to derive some correlations between the computational data obtained from ionic nitramine salts and their electric spark sensitivity values. Zhi et al. [361] used the lowest unoccupied molecular orbital energy and Mulliken charges of the nitro group, as well as the number of the aromatic rings and certain substituents on polynitroaromatic compounds to predict their electric spark sensitivity. Yan and Zeman [285] as well as Zeman and Jungová [286] have reviewed some predictive methods for the prediction of the electric spark sensitivity of some classes of explosives.

7.3 Simple methods for predicting electrostatic spark sensitivity based on the RDAD instrument

There are several simple methods for predicting the electric spark sensitivity of poly-nitroaromatic and nitramine explosives based on the RDAD instrument, which are described in the following sections.

7.3.1 Polynitroaromatic compounds

For polynitroaromatic compounds, the following equation can be used to predict the electric spark sensitivity from structural parameters as [362]:

$$E_{ES} = 4.60 - 0.733a + 0.724d + 9.16(b/d) - 5.14C_{R,OR}, \tag{7.2}$$

where E_{ES} is in J; $C_{R,OR}$ represents the presence of alkyl (−R) or alkoxy (−OR) groups according to the following conditions:
(1) The value of $C_{R,OR}$ is 1.0 for alkyl groups attached to a nitroaromatic ring, e. g. 2-methyl-1,3,5-trinitrobenzene.
(2) The value of $C_{R,OR}$ equals −2.0 for the attachment of an alkoxy group to a nitro-aromatic ring, e. g. 2-methoxy-1,3,5-trinitrobenzene.

Example 7.1. The use of equation (7.2) for 2-methoxy-1,3,5-trinitrobenzene (TNA) with the following molecular structure

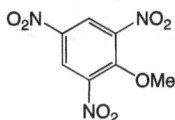

gives

$$E_{ES} = 4.60 - 0.733a + 0.724d + 9.16(b/d) - 5.14C_{R,OR}$$
$$= 4.60 - 0.733(7) + 0.724(7) + 9.16(5/7) - 5.14(-2)$$
$$= 21.36 \, J.$$

The measured value of E_{ES} is 28.59 J [348].

7.3.2 Cyclic and acyclic nitramines

It was found that the ratio of carbon to oxygen atoms, the presence of methylenenitramine units ($-CH_2NNO_2-$) in cyclic nitramines, as well as $-COO-$ (or amide) groups can be used to predict the electrostatic sensitivity of cyclic and acyclic nitramines as follows [363]:

$$E_{ES} = 3.460 + 6.504(a/d) - 4.059C_{CH_2NNO_2 \geq 3, C(=O)(O \text{ or } NH)}, \tag{7.3}$$

where $C_{CH_2NNO_2 \geq 3, C(=O)(O \text{ or } NH)}$ is either the number of methylenenitramine groups greater than two in cyclic nitramines or the presence of $-COO-$ (or $-CONH-$) functional groups.

Example 7.2. If equation (7.3) is used for 2,4,5-trinitro-2,4,6-triazaheptane (ORDX) with the following molecular structure,

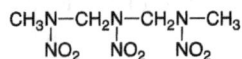

CH₃N—CH₂N—CH₂N—CH₃
  | | |
  NO₂ NO₂ NO₂

it gives

$$E_{ES} = 3.460 + 6.504(a/d) - 4.059C_{CH_2NNO_2 > 2, C(=O)(O \text{ or } NH)}$$
$$= 3.460 + 6.504(4/6) - 4.059(0)$$
$$= 7.80 \, J.$$

The measured value of E_{ES} equals 8.08 J [364].

7.3.3 General correlation for polynitroaromatics as well as cyclic and acyclic nitramines

For various nitroaromatic and nitramine compounds, it was shown that the following correlation can be used to prediction the electric spark sensitivity [284]:

$$E_{ES} = 5.12 + 2.323\left(\frac{a}{d}\right) + 1.513\left(\frac{b}{d}\right) + 7.519E_{ES}^+ - 3.637E_{ES}^-, \tag{7.4}$$

where E_{ES}^{+} and E_{ES}^{-} are correcting functions that can increase and decrease the predicted results, based on the ratios of the number of carbon and hydrogen to oxygen atoms, respectively. The values of E_{ES}^{+} and E_{ES}^{-} can be given as

(1) Prediction of E_{ES}^{+}:
   (a) For nitroaromatic energetic compounds, the values of E_{ES}^{+} are 2.0, 1.0 and 0.75 in the presence of $-OR$, three $-NH_2$ and two $-OH$ groups, respectively;
   (b) For nitramines, if the $-COO-$, $C-O-C$ or tertiary amine $(NR_1R_2R_3)$ groups are present, the value of E_{ES}^{+} is 0.75;
   (c) For both nitroaromatics and nitramines, E_{ES}^{+} is 1.0 in the presence of the amide group $(-NH-OC-)$.

(2) Prediction of E_{ES}^{-}: The value of E_{ES}^{-} equals 1.0 for:
   (a) the attachment of only one CH_x- or $Ar-$ to an aromatic ring in the case of nitroaromatic compounds;
   (b) if the number of methylenenitramine (CH_2NNO_2) moieties is greater than, or equal to three in cyclic nitramines (except for cage nitramines), e. g. 2,4,6,8,10,12-hexanitro-2,4,6,8,10,12-hexaazaisowurtzitane (HNIW).

Example 7.3. The use of equation (7.4) for both compounds given in previous examples gives

$$E_{ES} = 5.12 + 2.323\left(\frac{a}{d}\right) + 1.513\left(\frac{b}{d}\right) + 7.519E_{ES}^{+} - 3.637E_{ES}^{-}$$

TNA:

$$= 5.12 + 2.323\left(\frac{7}{7}\right) + 1.513\left(\frac{5}{7}\right) + 7.519(2) - 3.637(0)$$

$$= 23.56\,J$$

ORDX:

$$= 5.12 + 2.323\left(\frac{4}{6}\right) + 1.513\left(\frac{10}{6}\right) + 7.519(0) - 3.637(0)$$

$$= 9.19\,J.$$

Thus, the deviation of the predicted results of TNA and ORDX from the measured values are 5.03 and $-1.11\,J$, respectively. Meanwhile, the use of the complex quantum mechanical method of Wang et al. [353, 354] gives 6.01 J (Dev $=$ 22.58 J) and 9.09 J (Dev $= -1.01$ J) for TNA and ORDX, respectively.

7.4 Simple prediction of electrostatic spark sensitivity based on the new ESZ KTTV instrument

Since the new instrument of ESZ KTTV gives more reliable experimental data than the old system RDAD, several attempts have been done to introduce the improved correlations based on ESZ KTTV instrument. Moreover, several correlations have also been introduced to extend the output of previous works based on the RDAD instrument to the ESZ KTTV system. These methods are demonstrated in the following sections.

7.4.1 Polynitro arenes based on ESZ KTTV

A suitable correlation has been introduced to estimate the electrostatic sensitivity of polynitro arenes based on ESZ KTTV as follows [365]:

$$E_{ES,PNA}(ESZ\ KTTV) = 220.6 - 7.91d - 71.08f + 191.4E^+_{ES,PNA}, \tag{7.5}$$

where $E_{ES,PNA}(ESZ\ KTTV)$ is the electric spark sensitivity of polynitro arenes based on the ESZ KTTV system in mJ; d and f are equal to the number of moles of oxygen and chlorine atoms, respectively; $E^+_{ES,PNA}$ is a correcting function that increases the predicted results based on d and f. The value of $E^+_{ES,PNA}$ is specified according to the following conditions:

(a) The presence existence of molecular moieties and as well as direct attachment of three alkyl groups or one aromatic ring to another aromatic ring and dinitrobenzene: The value of $E^+_{ES,PNA}$ equals 0.7 for the presence of one of the mentioned molecular moieties.

(b) The presence of three $-NH_2$ groups: The value $E^+_{ES,PNA}$ equals 0.5.

(c) The presence of group: The value of $E^+_{ES,PNA}$ is equal to 0.8.

(d) The presence of : The value of $E^+_{ES,PNA}$ equals 1.5.

Example 7.4. The use of equation (7.5) and condition (c) for 1,3,7,9-tetranitro-10H-phenothiazine 5,5-dioxide (TNPTD) with the following molecular structure

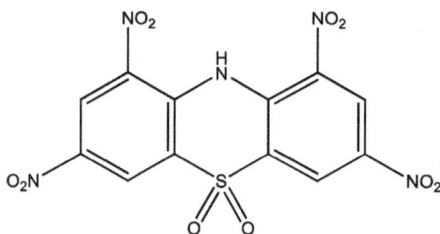

gives:

$$E_{ES,PNA}(ESZ\ KTTV) = 220.6 - 7.91d - 71.08f + 191.4E^+_{ES,PNA}$$
$$= 220.6 - 7.91 \times 10 - 71.08 \times 0 + 191.4 \times 0.8 = 294.6\ mJ$$

The measured electrostatic sensitivity of TNPTD using ESZ KTTV instrument is 363.3 mJ [366].

It is possible to correlate the predicted results of the RDAD for the energetic compounds given in Section 7.3 to the ESZ KTTV instrument because most of the available predictive methods [367] and the new software EMDB [69] can estimate the electric spark sensitivity based on RDAD instrument. It was shown that the predicted results of the EMDB [69] can be used to find electric spark sensitivity based on ESZ KTTV instrument as follows [365]:

$$E_{ES,PNA}(ESZ\ KTTV) = 72.5 + 10.40E_{ES,PNA}(RDAD) + 139.4E^{+'}_{ES,PNA} - 62.03E^{-}_{ES,PNA}, \quad (7.6)$$

where $E_{ES,PNA}(RDAD)$ is the electrostatic sensitivity based on the RDAD system for polynitro arenes in mJ; $E^{+'}_{ES,PNA}$ and $E^{-}_{ES,PNA}$ are two correcting functions that can increase and decrease the predicted results based on the RDAD system, respectively. The value of $E^{+'}_{ES,PNA}$ equals 1.0 for polynitro arenes that follow the following conditions:

(a) The presence of ⟨structure⟩ and direct attachment of three alkyl groups to the aromatic ring as well as dinitrobenzene.
(b) Direct attachment of picryl group to triazene ring or to another picryl group.

The value of $E^{-}_{ES,PNA}$ is also equal to 1.0 for the presence of $-N=N-$ or $-Cl$ groups.

Example 7.5. The use of equation (7.6) and the experimental value of $E_{ES,PNA}(RDAD) = 13.37\,J$ for (E)-bis(3-methyl-2,4,6-trinitrophenyl)-diazene (DMHNAB) with the following molecular structure

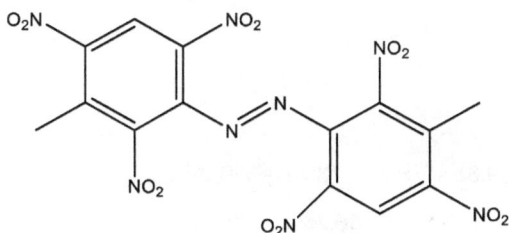

provides:

$$E_{ES,PNA}(ESZ\ KTTV) = 72.5 + 10.40E_{ES,PNA}(RDAD) + 139.4E^{+'}_{ES,PNA} - 62.03E^{-}_{ES,PNA}$$
$$= 72.5 + 10.40 \times 13.37 + 139.4 \times 0 - 62.03 \times 1.0 = 149.5\,mJ$$

The reported electrostatic sensitivity for DMHNAB using ESZ KTTV instrument is 118.2 mJ [366].

7.4.2 Nitramines based on ESZ KTTV

A reliable correlation has been introduced for assessment of the electric spark sensitivity of nitramines based on the ESZ KTTV as follows [368]:

$$E_{ES,NTA}(ESZ\ KTTV) = 110.7 + 129.4a - 39.8b + 66.2c - 78.7d + 350.0E^+_{ES,NTA}, \quad (7.7)$$

where $E_{ES,NTA}$ (ESZ KTTV) is the electric spark sensitivity of nitramines based on the ESZ KTTV system in mJ; a, b, c, and d are equal to the number of moles of carbon, hydrogen, nitrogen, and oxygen atoms, respectively; $E^+_{ES,NTA}$ is a correcting function that increases the predicted results based on elemental composition. The value of $E^+_{ES,NTA}$ is specified according to the following conditions:

(a) The presence of $-N(NO_2)-CH_2-CH_2-N(NO_2)-$ per cycle in cyclic or acyclic nitramines: The value of $E^+_{ES,NTA}$ equals 0.5.
(b) The presence of two $-N(NO_2)-CH_2-$ in cyclic nitramines with less than six-membered ring: The value of $E^+_{ES,NTA}$ is equal to 1.0.

Example 7.6. The use of equation (7.7) and condition (a) for 1,4,5,8-tetranitrodecahydropyrazino[2,3-b]pyrazine (TNAD) with the following molecular structure

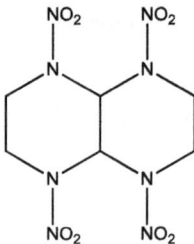

gives:

$$E_{ES,NTA}(ESZ\ KTTV) = 110.7 + 129.4a - 39.8b + 66.2c - 78.7d + 350.0E^+_{ES,NTA}$$

$$= 110.7 + 129.4 \times 6 - 39.8 \times 10 + 66.2 \times 8 - 78.7 \times 8 + 350.0 \times 0.5$$

$$[2pt] = 564.1\ mJ$$

The measured electrostatic sensitivity of TNAD using ESZ KTTV instrument is 520.0 mJ [369].

It is important to have a suitable correlation for conversion of the reported data based on RDAD to the ESZ KTTV, which can be done with the following equation by considering some specific molecular fragments:

$$E_{ES,NTA}(ESZ\ KTTV) = 616.8 - 27.80E_{ES,NTA}(RDAD) - 296.1E^-_{ES,NTA} \quad (7.8)$$

where $E_{ES,NTA}(RDAD)$ is the electric spark sensitivity based on the RDAD system in mJ; $E^-_{ES,NTA}$ is a correcting function that decreases the predicted results based on the RDAD system. The value of $E^-_{ES,NTA}$ equals 1.0 for cyclic nitramines with the equal number of $-CH_2-$ and $-N(NO_2)-$ groups in more than five-membered ring as well as cyclic nitramines containing $>C(NO_2)_2$ group.

Example 7.7. The use of equation (7.8) and the measured value of $E_{ES,NTA}(RDAD) = 2.96\,J$ for 11,3,5,7,9-pentanitro-1,3,5,7,9-pentazecane (DECAGEN) with the following molecular structure

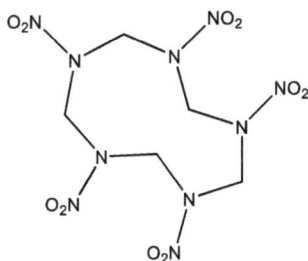

gives:

$$E_{ES,NTA}(ESZ\ KTTV) = 616.8 - 27.80E_{ES,NTA}(RDAD) - 296.1E^-_{ES,NTA}$$
$$= 616.8 - 27.80 \times 2.96 - 296.1 \times 1.0 = 238.4\,mJ$$

The measured electrostatic sensitivity of DECAGEN using ESZ KTTV instrument is 276.5 mJ [369].

7.4.3 Quaternary ammonium-based energetic ionic liquids or salts based on ESZ KTTV

A simple method has been introduced for the calculation of electrostatic sensitivity of quaternary ammonium-based energetic ionic liquids or salts based on ESZ KTTV as follows [370]:

$$ES_{IL}\ (mJ) = 635 + 725a_{cat} - 241c_{cat} - 409d_{cat} + 117a_{ani} + 689ES^+_{IL} - 551ES^-_{IL}, \quad (7.9)$$

where ES_{IL} (mJ) is the sensitivity toward the electrical discharge of a desired energetic ionic compound in mJ; a_{cat}, c_{cat} and d_{cat} are the number of carbon, nitrogen, and oxygen atoms in cation, respectively; a_{ani} is the number of carbon atoms in the anion. Two parameters ES^+_{IL} and ES^-_{IL} are two correcting functions. Tables 7.1 and 7.2 show different types of cations and anions for which the values of ES^+_{IL} and ES^-_{IL} equal 1.0.

Table 7.1: Different types of cations and anions for which the value of ES_{IL}^+ is 1.0.

Cation	Anion

Table 7.2: Different types of cations and anions for which the value of ES_{IL}^- is 1.0.

Cation	Anion

Example 7.8. The use of equation (7.9) for the following compound

gives:

$$ES_{IL} \text{ (mJ)} = 635 + 725a_{cat} - 241c_{cat} - 409d_{cat} + 117a_{ani} + 689ES_{IL}^{+} - 551ES_{IL}^{-}$$
$$= 635 + 725 \times 1 \times 2 - 241 \times 3 \times 2 - 409 \times 0 + 117 \times 4 + 689 \times 0 - 551 \times 1.0$$
$$= 556 \text{ mJ}$$

The measured value of ES_{IL} for this compound is 750 mJ [371].

7.5 Some aspects of predictive methods

The currently available predictive methods cannot predict electrostatic sensitivity versus the grain size because the electric spark sensitivity depends on the size and shape of crystals. For some isomers, the difference in the sensitivities may be large, and this can be attributed to the different behavior of nitro groups in different positions, e. g. electric spark sensitivity for 1,3-dinitrobenzene and 1,4-dinitrobenzene are 3.15 and 18.38 J [348], respectively. Since experimental data of the electric spark sensitivity of different isomers is rare, use of the predictive methods which are available may result in large deviations for some isomers. Fortunately, equations (7.2) to (7.4) can predict the electric spark sensitivity which is close to the average value. For example, equation (7.4) predicts a value of 10.11 J for both 1,3-dinitrobenzene and 1,4-dinitrobenzene and this is close to the average value of the measured values for the two isomers.

7.6 Summary

Some developments for the prediction of the electric spark sensitivity of secondary explosives have been reviewed in this chapter. Since different factors can influence the sensitivity resulting from different stimuli, the main intent in this work was to illustrate the best available simple methods to evaluate the electrostatic sensitivity of energetic compounds. Three simple equations (7.2), (7.3) and (7.4) can be can be easily used to theoretically predict the magnitude of the electrostatic sensitivity of based on the RDAD instrument polynitroaromatics, nitramines, and also of both nitroaromatics and nitramines, respectively. Thus, these equations are useful models in terms of their accuracy and simplicity because they require only knowledge of the molecular structure of energetic compounds, which is always known. Due to the different behavior – in terms of the electric spark sensitivity – of nitramines and nitroaromatic compounds, equation (7.4) has the advantage that it can be applied to nitroaromatic compounds which contain $-N-NO_2$ groups, e. g. 1-(methylnitramino)-2,4,6-trinitrobenzene (TETRYL). Since the new instrument of ESZ KTTV gives more reliable experimental data than the old system RDAD, equations (7.5) and (7.7) provide two correlations for the prediction of the electrostatic sensitivity based on the ESZ KTTV instrument of polynitroaromatics and nitramines, respectively. Moreover, equations

(7.6) and (7.8) give a suitable pathway for conversion of the outputs of software EMDB [69] based on the RDAD to those based on the ESZ KTTV for polynitroaromatics and nitramines, respectively. Equation (7.9) can also be used to estimate the electrostatic sensitivity of quaternary ammonium-based energetic ionic liquids or salts based on ESZ KTTV.

8 Shock sensitivity

Gap test data is useful to indicate the shock sensitivity of an explosive, and it is nowadays widely used to determine the shock sensitivity of a desired explosive. Different gap tests have been used to qualitatively measure the shock wave amplitude which is required to initiate detonation in explosives, e. g. at Naval Surface Warfare Center (NSWC) and Los Alamos National Laboratory (LANL). Large scale and small scale gap tests are two convenient methods for measuring shock sensitivity [336]. In contrast to the results of impact sensitivity – which are often not reproducible because factors in the impact experiment that might affect the formation and growth of hot spots can strongly affect the measurements – reliable shock sensitivity tests exist. Furthermore, the reported data of impact sensitivities are extremely sensitive to the conditions under which the tests are performed.

There are several reviews in which different methods for predicting the shock sensitivity of different pure and composite explosives have been considered [282, 284, 286]. Price [372] has studied different important factors in shock wave sensitivity tests. For five explosives with closely related structures, i. e. TNB, DIPAM, MATB, DATB and TATB, Storm et al. [336] have shown that there is a linear correlation between the impact and shock sensitivity under specified conditions. Due to the dependence of the results of impact sensitivity tests on the conditions of the experiment, they used the impact sensitivity as measured at LANL and/or NSWC using the Bruceton method, type 12 tools, 2.5 kg weight, 40 mg sample, 5/0 sand paper and 25 trials. For seven polynitroaromatics, Owen et al. [304] also found that the measured impact and shock sensitivities can correlate with an approximation of the electronegativity potential at the midpoint of the C–N bond for the longest C–NO$_2$ bond in each molecule. Tan et al. [373] used quantum mechanical calculations (DFT/BLYP/DNP) to calculate the bond dissociation energies of X–NO$_2$ (X = C, N, O) and Mulliken charges of nitro groups for 14 examples of nitro compounds. Among the different approaches for the prediction of the shock sensitivity, there are two simple methods based on simple structural parameters, which are discussed in this chapter.

8.1 Small-scale gap test

Studies of the shock sensitivity as measured using the NSWC small scale gap test shows that three special structural parameters may affect their values including:
(1) the distribution of oxygen between carbon and hydrogen;
(2) the existence of nitramine groups or a α-CH linkage in nitroaromatic compounds;

https://doi.org/10.1515/9783110740158-008

(3) the difference in the number of amino and nitro groups in aminoaromatic (Ar–NH$_2$) energetic compounds.

Thus, the following general equation can be applied to C$_a$H$_b$N$_c$O$_d$ explosives [374]:

$$P_{90\%\text{ TMD}} = 16.79 + 2.262(a + b/2 - d) - 6.314E^0_{\alpha\text{CH/NNO}_2}$$
$$+ 17.72(1.93n_{\text{NH}_2} - n_{\text{NO}_2})_{\text{pure}}, \tag{8.1}$$

$$P_{95\%\text{ TMD}} = 21.96 + 2.479(a + b/2 - d) - 6.3677E^0_{\alpha\text{CH/NNO}_2}$$
$$+ 32.92(1.93n_{\text{NH}_2} - n_{\text{NO}_2})_{\text{pure}}, \tag{8.2}$$

$$P_{98\%\text{ TMD}} = 25.45 + 2.211(a + b/2 - d) - 4.162E^0_{\alpha\text{CH/NNO}_2}$$
$$+ 46.39(1.93n_{\text{NH}_2} - n_{\text{NO}_2})_{\text{pure}}, \tag{8.3}$$

where $P_{90\%\text{ TMD}}$, $P_{95\%\text{ TMD}}$ and $P_{98\%\text{ TMD}}$ are the pressures in kbar which are required to initiate material pressed to 90 %, 95 % and 98 % of theoretical maximum density (TMD); $a + b/2 - d$ is a parameter that shows the distribution of oxygen between carbon and hydrogen; $E^0_{\alpha\text{CH/NNO}_2}$ is a parameter that indicates the existence of a α-CH linkage in nitroaromatic compounds, or a N–NO$_2$ functional group; $(1.93n_{\text{NH}_2} - n_{\text{NO}_2})_{\text{pure}}$ is the difference in the number of amino and nitro groups in aminoaromatic energetic compounds when $1.93n_{\text{NH}_2} \geq n_{\text{NO}_2}$. The value of $E^0_{\alpha\text{CH/NNO}_2}$ equals 1.0 for nitramines or for the presence of a α-CH linkage in nitroaromatic compounds, e. g. TNT. It is also equal to 1.0 for composite explosives containing more than or equal to 50 % of nitramines or α-C–H linkage in nitroaromatic compounds. The parameter $a + b/2 - d$ may affect the sensitivity of different classes of explosives. Since the presence of the N–NO$_2$ functional group and α-CH linkage in nitroaromatic compounds can increase the sensitivity of these compounds, the coefficient of $E^0_{\alpha\text{CH/NNO}_2}$ has a minus sign. The attachment of amino groups to an aromatic ring may enhance the stability of an energetic compound, while the addition of electron withdrawing groups (such as NO$_2$ groups) leads to a reduction in the stabilization of the aromatic ring. Thus, amino groups partially counteract the electron withdrawing effect of nitro groups which enhances the stabilization of the aromatic ring through the parameter $(1.93n_{\text{NH}_2} - n_{\text{NO}_2})_{\text{pure}}$.

The deviation of the values obtained from equations (8.1) to (8.3) from the measured values becomes large for very fine particle sizes because small particle size can reduce the shock sensitivity at high density.

Example 8.1. The use of equations (8.1) to (8.3) to calculate the results of small-scale gap test of 3,3'-diamino-2,2',4,4',6,6'-hexanitrobiphenyl (C$_{12}$H$_6$N$_8$O$_{12}$) gives

$$P_{90\%\text{ TMD}} = 16.79 + 2.262(a + b/2 - d) - 6.314E^0_{\alpha\text{CH/NNO}_2}$$
$$+ 17.72(1.93n_{\text{NH}_2} - n_{\text{NO}_2})_{\text{pure}}$$
$$= 16.79 + 2.262(12 + 6/2 - 12) - 6.314(0) + 17.72(0)$$
$$= 23.58\text{ kbar},$$

$$P_{95\%\,TMD} = 21.96 + 2.479(a + b/2 - d) - 6.3677E^0_{\alpha CH/NNO_2}$$
$$+ 32.92(1.93n_{NH_2} - n_{NO_2})_{pure}$$
$$= 21.96 + 2.479(12 + 6/2 - 12) - 6.3677(0) + 32.92(0)$$
$$= 29.40\,\text{kbar},$$
$$P_{98\%\,TMD} = 25.45 + 2.211(a + b/2 - d) - 4.162E^0_{\alpha CH/NNO_2}$$
$$+ 46.39(1.93n_{NH_2} - n_{NO_2})_{pure}$$
$$= 25.45 + 2.211(12 + 6/2 - 12) - 4.162(0) + 46.39(0)$$
$$= 32.08\,\text{kbar}.$$

The measured values of $P_{90\%\,TMD}$, $P_{95\%\,TMD}$ and $P_{98\%\,TMD}$ are 25.11, 29.71 and 33.04 kbar [336], respectively.

8.2 Large-scale gap test

Different gap tests were used to qualitatively measure the shock wave amplitude required to initiate detonation in an explosive. In the large-scale gap test, a shock pressure of uniform magnitude is produced by a detonating charge of high explosive, which is transmitted to the test explosive through an attenuating inert barrier or gap. Since the thickness of the barrier between the donor and test (acceptor) explosives can be varied, one can determine the barrier thickness required to inhibit detonation in the test explosive half of the time (G_{50}). Two test configurations were used by LANL, in which the diameter of the cylinder acceptor charge in the small-scale test is 12.7 mm and in the large-scale is 41.3 mm [375]. An explosive in the small-scale test, whose detonation failure diameter is near to or greater than the diameter of the acceptor charge cannot be tested. The large-scale has an advantage over the small-scale it can be tested in this situation. For the large-scale, the test method is to fire a few preliminary shots to determine the spacer thickness that allows detonation in the test explosive. When the shots are fired with the spacer thickness being alternately increased and decreased, the spacer thickness that allows a 50 % detonation probability in the acceptor explosive is determined. The dent produced in a witness plate can ascertain detonation of the acceptor charge. Thus, a deep defined dent in the steel witness plate shows detonation of the test explosive has occurred.

Large-scale shock sensitivities of various explosives rely on physical and chemical structural parameters, and for various explosives depend on four main essential parameters:
(1) initial density;
(2) percent void;
(3) distribution of oxygen between carbon and hydrogen;
(4) structural parameter C–N(NO$_2$)–C for pure nitramine explosives.

Pure nitramines containing the C–N(NO$_2$)–C linkage are more sensitive than other pure explosives containing only the C–NO$_2$ linkage. The following general equation can be used for the prediction of the large-scale shock sensitivities of various types of C$_a$H$_b$N$_c$O$_d$ pure and mixed explosives [376]:

$$G_{50} = 171.47 - 69.10\rho_0 - 2.61(a + b/2 - d) - 0.961Void_{\text{theo}}$$
$$+ 12.32(\text{C–N(NO}_2)\text{–C})_{\text{pure}}, \tag{8.4}$$

where G_{50} is in mm, $Void_{\text{theo}}$ is theoretical calculated percent of voids that can be obtained from

$$\frac{(1/\rho_0 - 1/\rho_{\text{TM}})}{1/\rho_0} \times 100,$$

where ρ_{TM} is the theoretical maximum density.

Equation (8.4) can be applied to pure and composite mixtures that are prepared under vacuum cast, cast, hot-pressed and pressed conditions because deviations may be large for creamed, granular and flake situations.

Example 8.2. PBX-9007 has the composition 90/9.1/0.5/0.4 RDX/Polystyrene/DOP/Rosin (C$_{1.97}$H$_{3.22}$N$_{2.43}$O$_{2.44}$). If the values of ρ_0 and ρ_{TM} are 1.646 and 1.697 g/cm^3, respectively, the use of equation (8.4) gives

$$Void_{\text{theo}} = \frac{(1/\rho_0 - 1/\rho_{\text{TM}})}{1/\rho_0} \times 100 = \frac{(1/1.646 - 1/1.697)}{1/1.646} \times 100 = 3.005$$

$$G_{50} = 171.47 - 69.10\rho_0 - 2.61(a + b/2 - d)$$
$$- 0.961Void_{\text{theo}} + 12.32(\text{C–N(NO}_2)\text{–C})_{\text{pure}}$$
$$= 171.47 - 69.10(1.646) - 2.61(1.97 + 3.22/2 - 2.44)$$
$$- 0.961(3.005) + 12.32(0)$$
$$= 51.87 \, \text{mm}.$$

The measured values of G_{50} for PBX-9007 is 52.91 mm [375].

8.3 Critical diameter of solid pure and composite high explosives

The critical (failure) diameter is the minimum diameter of a cylindrical charge of a high explosive that sustains a high order steady-state detonation [377], which depends on confinement, particle size, and initial temperature of the sample [377, 378]. The critical diameters of primary and high explosives are usually very small sometimes in the μm and in the range mm to cm region, respectively. Moreover, the experimental values of critical diameters of explosives depend on the conditions in which they are confined, e. g. the critical diameters of 15 and 7 mm were reported for TNT (cast) with loading density about $1.6\,\text{g cm}^{-3}$ encased in $0.2\,\text{mm}$ paper at 291 K and 290 K after one temperature cycle to 77 K, respectively [377].

Determination of critical diameters of high explosives with high detonability can be difficult because it is necessary to manufacture high-density charges of diameter less than 1 mm [379]. It is unfeasible to determine the critical diameter of high explosives with low detonability (charges of large diameter > 100 mm) under laboratory conditions because it is necessary to investigate them [379]. The charges with higher detonability require a smaller critical diameter [380]. Some attempts have been developed to use suitable predictive methods for reducing the number of experiments of pure and composite explosives. A complex critical-diameter theory based on complex variables as input parameters has been developed for some specific high explosives [379, 381, 382]. For the prediction of the critical diameter of a high-explosive charge under shock-wave compression, Kobylkin [379] has shown that it is necessary to know the shock adiabat, detonation velocity, and the generalized kinetic characteristic of decomposition. The generalized kinetic characteristic of decomposition can be found from the measurement of the shock-wave amplitude on the distance the shock wave that travels during shock-wave initiation of the high explosive charge. A simple method has been introduced for calculating critical diameter under the unconfined condition as follows [383]:

$$d_c = -3.19c + 5.38d + 21.80C_{\text{Shock}}, \tag{8.5}$$

where d_c is the critical diameter in mm; c and d are the numbers of moles of nitrogen and oxygen atoms, respectively; C_{Shock} is a correcting function. For pure and composite explosives, the following situations should be considered for evaluation of C_{Shock}:

(1) *Shock sensitivity and impact sensitivity*: The shock sensitivity of explosives is more important in its safety assessment as compared to the impact sensitivity [336]. Equation (8.3) was used to specify the contribution of the shock sensitivity of both pure and composite explosives. Equations (6.11) and (6.12) are used with the symbol E_{IS} in J unit, i. e. $E_{\text{IS}} = 0.245H_{50}$, to mention the contribution of impact sensitivity of pure explosives.

(2) *Energetic compounds with restrictions of c and d*: As seen in equation (8.5), the signs of c and d are negative and positive, respectively. Thus, the ratio of c/d should be less than 1.18.

According to the above situations, different values of C_{Shock} are given as:

(a) *Pure explosives* – Several situations are considered:
   (i) The value of C_{Shock} equals -0.40 if $P_{98\%\,\text{TMD}} > 38\,\text{kbar}$ and $E_{\text{IS}} > 28\,\text{J}$.
   (ii) For $E_{\text{IS}} < 22\,\text{J}$, the fixed value of critical diameter should be considered, i. e. $d_c = 2.5$.

(b) *Composite explosives* – The assessment of C_{Shock} can be done as follows:

(iii) For composite explosives including PETN, $C_{Shock} = -0.30$.

(iv) The value of C_{Shock} is equal to -0.10 if $P_{98\% \text{ TMD}} < 24.4$ kbar.

(c) *If $c/d > 1.18$: The value of C_{Shock} is 0.80.*

The process of casting can affect the value of d_c for the unconfined condition, e. g. the values of d_c are $22.0 < d_c < 25.4$ and $12.6 < d_c < 16.6$ mm poured as cloudy slurry and creamed for TNT, respectively [377]. Moreover, axially oriented TNT crystals give unstable detonation while radially oriented crystals detonate smoothly. The calculated data of equation (8.5) for cast explosives are more closed to the experimental data than pressed explosives. Two reasons can be introduced for this situation:

(1) The actual density of cast explosives is more closed to its theoretical maximum density than that of pressed explosives because there are fewer voids in cast explosives.

(2) Distributing of stresses in pressed charges exists substantially stronger than that in cast charges.

Example 8.3. z-TACOT (Tetranitro-2,3,5,6-dibenzo-1,3a,4,6a-tetraazapentalene) has the following molecular structure:

Chemical Formula: $C_{12}H_4N_8O_8$
Molecular Weight: 388.21

The use of equation (8.3), as well as equations (6.11) and (6.12), gives:

$$(\log H_{50})_{core} = -0.584 + 61.62a' + 21.53b' + 27.96c'$$
$$= -0.584 + (61.62 \times 12 + 21.53 \times 4 + 27.96 \times 8)/388.21 = 2.119$$

$$\log H_{50} = (\log H_{50})_{core} + 84.47\frac{F^+}{Mw} - 147.1\frac{F^-}{Mw}$$
$$= 2.119 + 84.47\frac{0}{388.21} - 147.1\frac{0}{388.21} = 2.119$$

$$H_{50} = 131 \text{ cm}$$

$$E_{IS} = 0.245H_{50} = 0.245 \times 131 = 32.2 \text{ J}$$

$$P_{98\% \text{ TMD}} = 25.45 + 2.211(a + b/2 - d) - 4.162E^0_{\alpha CH/NNO_2} + 46.39(1.93n_{NH_2} - n_{NO_2})_{pure}$$
$$= 25.45 + 2.211(12 + 4/2 - 12) - 4.162(0) + 46.39(0)$$
$$= 38.72 \text{ kbar}$$

The value of C_{Shock} equals -0.4 because $P_{98\% \, TMD} > 38$ kbar and $E_{IS} > 28$ J. Thus, equation (8.5) gives:

$$d_c = -3.19c + 5.38d + 21.80C_{Shock}$$
$$= -3.19 \times 8 + 5.38 \times 8 + 21.80(-0.4) = 8.80 \, \text{mm}$$

The measured value of d_c for z-TACOT is 3.0 mm [377].

Example 8.4. Composite explosive Comp A-3 (pressed) consisting of 91 % RDX and 9 % wax binder has the chemical formula $C_{1.87}H_{3.74}N_{2.46}O_{2.46}$. Since the presence of RDX containing nitramine group ≥ 50 %, $E^0_{\alpha CH/NNO_2} = 1.0$. The use of equation (8.3) gives:

$$P_{98\% \, TMD} = 25.45 + 2.211(a + b/2 - d) - 4.162E^0_{\alpha CH/NNO_2} + 46.39(1.93n_{NH_2} - n_{NO_2})_{pure}$$
$$= 25.45 + 2.211(1.87 + 3.74/2 - 2.46) - 4.162(1.0) + 46.39(0)$$
$$= 24.12 \, \text{kbar}$$

The value of C_{Shock} is equal to -0.10 because of $P_{98\% \, TMD} < 24.4$ kbar. Thus, equation (8.5) gives:

$$d_c = -3.19c + 5.38d + 21.80C_{Shock}$$
$$= -3.19 \times 2.46 + 5.38 \times 2.46 + 21.80(-0.1) = 3.21 \, \text{mm}$$

The measured value of d_c for Comp A-3 is 2.2 mm [377].

8.4 Summary

This chapter introduced different approaches for the predicting shock sensitivity of pure and composite explosives using small- and large-scale gap tests. The simple equations (8.1) to (8.4) have two major advantages with respect to the impact sensitivity correlations given in Chapter 6, and which are:
(1) since a high percentage of errors are usually attributed to the reported experimental measurements from different sources for impact sensitivities, there is a large uncertainty in the different methods of impact sensitivity predictions as compared to equations (8.1) to (8.4) for small and large-scale gap thickness shock sensitivity;
(2) different correlations for the impact sensitivity can only be applied for pure explosives, but equations (8.1) to (8.4) can be used for both pure and composite explosives. Equation (8.5) can be used to find critical diameter of pure and composite explosives by considering equation (8.3) as well as equations (6.11) and (6.12).

9 Friction sensitivity

Friction is one of the stimuli that can cause explosions and fires in pyrotechnic compositions and explosives [285]. It is important to investigate important aspects of friction sensitivity because it shows the behavior of an energetic compound with respect to friction stimuli. The BAM large friction tester can be used to determine the friction sensitivity of a sample, in which approximately 30 mg of the sample is placed on a porcelain plate [384]. Therefore, the BAM friction tester is widely accepted as a standard friction tester in Europe. The surfaces of both the porcelain plates and pegs have uniform roughness, so that the porcelain pin is lowered onto the sample and a weight is placed on the arm to produce the desired load. The tester is activated and the porcelain plate is moved once forward and backwards. The results of friction sensitivity are observed as either a reaction in form of a flash, smoke, and/or audible report, or no reaction.

In contrast to impact, electric spark, and shock sensitivity, friction sensitivity does not attract the attention of theoretical chemists/physicists because of the shortcomings of the influence of the results of friction sensitivity on experimental data. Zeman and Jungová [286] have reviewed several methods that have been used to predict the friction sensitivity of energetic materials. Jungová et al. [385] indicated that there is a relationship between the friction sensitivity of nitramines and their thermal decomposition parameters. Friedl et al. [386] showed the relationship between the friction sensitivity and surface electronic potentials of nitramines. Jungová et al. [387] also compared the friction sensitivity of nitramines with their impact sensitivities and heats of fusion. For some classes of nitramines [388], there is a semi logarithmic relationship between the impact and friction sensitivities. For the safe handling of energetic compounds, knowledge of the friction sensitivity may be important because friction is frequently encountered during mixing, pouring, sieving, priming and consolidation operations [389]. Two empirical methods are introduced for assessment of friction sensitivity of nitramines and quaternary ammonium-based energetic ionic liquids.

9.1 Friction sensitivity of nitramines

For nitramines, BAM friction or Julius Peters friction apparatus can be used to determine their friction sensitivity. In these experimental methods, the explosive sample is held between a porcelain plate and a porcelain peg under a given load. Frictional forces are applied by the horizontal movement of the porcelain plate. The relative sensitivity to friction is indicated by the lowest load which leads to ignition, crackling or explosion. It was shown that the friction sensitivity of nitramines can be related to

https://doi.org/10.1515/9783110740158-009

their molecular structure as [390]:

$$FS = 600.8 - \frac{2428.6b + 6481.4c + 9560.9d}{Mw} + 54.5P_{FS}^+ - 77.8P_{FS}^-, \tag{9.1}$$

where FS is the friction sensitivity in N; P_{FS}^+ and P_{FS}^- are two parameters that can be predicted on the basis of the molecular structure of the nitramines as follows.

(1) Acyclic nitramines: For those compounds containing more than two repetitive $[-CH_2N(NO_2)-]$ units, the value of P_{FS}^+ equals 1.0, whilst the value of P_{FS}^- is 0.5 for the presence of the $-N(NO_2)-CH_2-CH_2-N(NO_2)-$ molecular fragment. The value of P_{FS}^+ depends on the number of separate molecular $N(NO_2)-C=$ moieties ($n_{N(NO_2)-C=}$). It is equal to $P_{FS}^+ = 5 - 2n_{N(NO_2)-C=}$ except for $n_{N(NO_2)-C=} \geq 3$ in which $P_{FS}^+ = 0$.

(2) Cyclic nitramines: For cyclic nitramines containing equal numbers of nitramine and methylene groups, the value of P_{FS}^+ equals 1.0. For cyclic nitramines containing five membered rings, the values of P_{FS}^- are 0.75 and 1.25 for the presence of two and one $-N(NO_2)-$ groups per ring, respectively.

(3) The attachment of a $-N(NO_2)-$ group to a tetrazole ring: For 5-nitriminotetrazole salts, the values of P_{FS}^+ and P_{FS}^- are 0.5 in the presence of ammonium and hydroxyl-ammonium cations, respectively. The values of P_{FS}^- and P_{FS}^+ equal 1.25 and 0.5 for 5-nitriminotetrazole and its methyl derivatives, respectively.

Example 9.1. The use of equation (9.1) for the following energetic compound

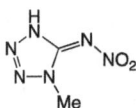

gives its friction sensitivity as

$$FS = 600.8 - \frac{2428.6b + 6481.4c + 9560.9d}{Mw} + 54.5P_{FS}^+ - 77.8P_{FS}^-$$

$$= 600.8 - \frac{2428.6(4) + 6481.4(6) + 9560.9(2)}{144.09} + 54.5(0.5) - 77.8(0)$$

$$= 158.0 \, N.$$

The measured friction sensitivity for this compound is 160 N [391].

9.2 Friction sensitivity of quaternary ammonium-based energetic ionic liquids

A simple correlation has been developed to predict friction sensitivity of quaternary ammonium-based energetic ionic liquids as follows [392]:

$$FS_{IL}(N) = 224 + 53.5a_{cat} - 25.9b_{cat} + 31.9c_{cat} + 134f_{cat} + 23.3a_{ani} + 162FS_{IL}^+ - 135FS_{IL}^-, \tag{9.2}$$

where $FS_{IL}(N)$ is friction sensitivity in newtons; a_{cat}, b_{cat}, c_{cat} and f_{cat} are the number of carbon, hydrogen, nitrogen and chlorine atoms in cation, respectively; a_{ani} is the number of carbon atoms in anion; FS_{IL}^{+} and FS_{IL}^{-} are two correcting functions. Tables 9.1 and 9.2 show different types of cations and anions for which values of FS_{IL}^{+} and FS_{IL}^{-} equal 1.0.

Table 9.1: Different types of cations and anions for which the value of FS_{IL}^{+} is 1.0.

Cation	Anion
	NO_3^-
	NO_3^- ClO_4^-
NH_4^+ 	

Example 9.2. The use of equation (9.2) for the following compound

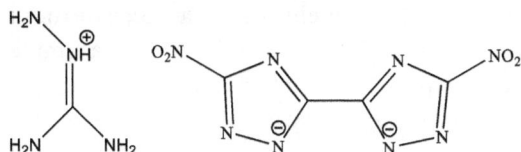

gives:

$$FS_{IL}(N) = 224 + 53.5a_{cat} - 25.9b_{cat} + 31.9c_{cat} + 134f_{cat} + 23.3a_{ani} + 162FS_{IL}^{+} - 135FS_{IL}^{-}$$
$$= 224 + 53.5 \times 1 \times 2 - 25.9 \times 7 \times 2 + 31.9 \times 4 \times 2 + 134 \times 0 + 23.3 \times 4 + 162 \times 0$$
$$- 135 \times 0 = 317\,N$$

The measured value of FS_{IL} for this compound is 360 N [393].

Table 9.2: Different types of cations and anions for which the value of FS_{IL}^- is 1.0.

Cation	Anion
	ClO_4^-
	NO_3^-
NH_4^+ 	

9.3 Summary

This chapter introduced a simple but reliable correlation between the friction sensitivity of nitramines and their molecular structures, which may be interesting for chemists and chemical industry. Equation (9.1) assumes that the friction sensitivity of a nitramine with the general formula $C_a H_b N_c O_d$ can be expressed as a function of its elemental composition and structural parameters. Since equation (9.1) confirms that the initiation reactivity of energetic compounds is intimately related to their chemical character and molecular structures, it can help to elucidate the mechanism of initiation in energetic materials by friction stimuli. Equation (9.2) develops a simple method for prediction of friction sensitivity of quaternary ammonium-based energetic ionic liquids based on the contributions of some atoms in cations and anions as well as two correcting functions.

10 Heat sensitivity

An ideal energetic compound shows high performance, low sensitivity and a good shelf life. The heat sensitivity and thermal stability of energetic materials are two important features in their shelf life and safety because knowledge of these properties is important in order to avoid undesirable decomposition or self-initiation during their handling, storage and application. Prediction of the thermal stability of a desired compound is an important starting point in the evaluation of its stability.

The thermolysis of energetic compounds can be used to estimate their thermal stability [394, 395]. The thermal stability of an energetic compound can be measured by various types of thermal analysis and gasometry, or by a variety of methods based on thermal explosions [396–398]. Among the different experimental methods, differential thermal analysis (DTA), differential scanning calorimetry (DSC) and thermogravimetric analysis methods are the thermoanalyical methods that are most widely used to examine the kinetic parameters of the thermolysis of energetic materials. The Soviet manometric method (SMM) is a suitable isothermal manometric method for energetic compounds, and uses a glass-compensating manometer of the Bourdon type to examine the kinetics of the thermolysis of energetic materials in vacuum. The data obtained by this method can be used to obtain the Arrhenius parameters of the non-autocatalyzed thermal decomposition of energetic materials. Experimental data from DTA and DSC can be converted to SMM data if a relationship such as a calibration curve exists between the results of the DTA and DSC with the results of SMM [399–404]. Since various factors affect the experimental data of the activation energies in the thermolysis of energetic materials, there is no uniform classification of a large majority of results obtained in various laboratories all over the world. In this chapter, some of the different methods for predicting parameters related to the thermal stability and heat sensitivity of energetic compounds – such as the activation energy and deflagration temperature – are reviewed. Some attempts have been done

https://doi.org/10.1515/9783110740158-010

to estimate thermal decomposition kinetic parameters of some classes of energetic compounds using complex methods, e. g. energetic cocrystals by the artificial neural network model [405], and correlations of detonation parameters with activation energy for nitric esters [406]. Some of the simple correlations which are based on the molecular structures of different classes of energetic compounds are also demonstrated.

10.1 Thermal kinetics correlations

Thermoanalytical methods such as DTA, DSC and thermogravimetry (TGA) can be used to determine Arrhenius parameters. Experimental critical temperatures for explosives of a given size and geometry can be obtained by a variety of tests, such as isothermal cook-off, slow cook-off, one-liter cook-off, and the isothermal time-to-explosion (Henkin test), in which the explosive may be confined or unconfined. Yan and Zeman [285], as well as Zeman and Jungová [286] have reviewed some methods for predicting the kinetic parameters of some classes of explosives. Several approaches are given here.

In the thermolysis of nitramines, homolysis of the N–NO$_2$ bond is a primary step for secondary nitramines, whereas the homolysis of primary nitramines is a bimolecular autoprotolytic reaction [407]. Nitramine groups contribute strongly to the intermolecular potential in the crystalline state because the longest N–N bonds are responsible for the homolytic reactivity of nitramines [408]. There are some linear relationships between the activation energies of nitramine decomposition with the ^{15}N NMR chemical shifts of nitrogen atoms of nitramino groups [281], heats of detonation, or the electronic charges at nitrogen atoms of the nitramines [409, 410]. A Muliken population analysis of the electron densities using the DFT B3LYP/6-31G** method can be used to calculate the electronic charges at the nitrogen atoms of nitramines [410]. Crystal lattice energies of nitramines do not generally differ from those of polynitroaromatics [396]. The activation energy (which corresponds to the slope in the Kissinger relationship) [411] can be used to evaluate the results of nonisothermal differential thermal analysis. Zeman [412] has used the modified Evans–Polanyi–Semenov (E–P–S) equation to interpret the chemical micromechanism governing the initiation of the detonation of energetic materials. Zeman [412] has used the heat of explosion and activation energy of the low temperature thermal decomposition to obtain the modified E–P–S equation for energetic materials. There are several simple correlations for the prediction of the activation energy of the thermolysis of several important classes of energetic compounds on the basis of their molecular structures, which are discussed in the following sections.

10.1.1 Nitroparrafins

A suitable relationship has been derived to predict the activation energies for the low-temperature non-autocatalyzed thermolysis (E_a) of nitroparaffins and the results of SMM as [413]:

$$\ln E_a = \frac{0.4190a + 0.1793b + 1.1914d}{Mw} \times 100. \tag{10.1}$$

Example 10.1. The use of equation (10.1) for 2,2-dinitropropane $(C_3H_6N_2O_4)$ gives:

$$\begin{aligned}
\ln E_a &= \frac{0.4190a + 0.1793b + 1.1914d}{Mw} \times 100 \\
&= \frac{0.4190(3) + 0.1793(6) + 1.1914(4)}{134.09} \times 100 \\
&= 5.294 \\
E_a &= 199.1\,\text{kJ/mol.}
\end{aligned}$$

The measured value for this compound is 198.74 kJ/mol [414].

10.1.2 Nitramines

The elemental composition and structural parameters of nitramines can be used to derive a suitable correlation for predicting E_a as follows [415].

$$\ln E_a = 0.5385 + 0.8951(5.012 - 0.0367a + 0.0255b - 0.0304c + 0.0407d) \\ + 0.1698P_{>5} \tag{10.2}$$

where $P_{>5}$ is equal to 1.0 for cyclic nitramines that contain rings which are larger than five membered rings, as well as nitramine cages.

Example 10.2. 4,10-Dinitro-2,6,8,12-tetraoxa-4,10-diazaisowurtzitane (TEX) has the following molecular structure:

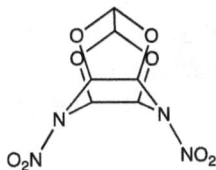

The use of equation (10.2) for TEX gives

$$\ln E_a = 0.5385 + 0.8951(5.012 - 0.0367a + 0.0255b - 0.0304c + 0.0407d) \\ + 0.1698P_{>5}$$

$$= 0.5385 + 0.8951(5.012 - 0.0367(6) + 0.0255(6) - 0.0304(4) + 0.0407(8))$$
$$+ 0.1698$$
$$= 5.317$$
$$E_a = 203.8 \text{ kJ/mol.}$$

The experimental value for TEX is 202.5 kJ/mol [407].

10.1.3 Polynitro arenes

It was found that in order to be able to predict the E_a of polynitro arenes several paramters are important, which include
(1) the contribution of the elemental composition;
(2) the number of $-NH(C=O)-C(=O)NH-$ groups (e. g. N,N′-bis (2,4,6-trinitrophenyl) oxamide) or more than one $-NH_2$ groups (e. g. 2,4,6-trinitrobenzene-1,3-diamine);
(3) the existence of either one α-CH (e. g. 2,4,6-trinitrotoluene), or methoxy group attached to one aromatic ring, or $CH_3O-[C(NO_2)-CH-C(NO_2)]$ – which has the opposite effect with respect to the second parameter.

The number of amino groups can affect the sensitivity and performance as well as physicothermal properties of nitroaromatic compounds [206]. Thus, the presence of amino groups can affect the value of the activation energy when more than one amino group is attached to aromatic rings. For polynitroaromatics with α-CH, it was indicated in previous chapters that their sensitivities with respect to different stimuli show different behaviors. It was found that the presence of one α-CH or $CH_3O-[C(NO_2)-CH-C(NO_2)]$ group in the molecular structure of polynitro arenes can influence the value of E_a. The value of E_a is a function of the aforementioned parameters, and which can be expressed by the following equation [416]:

$$\log(E_a) = 2.25 + 0.0337 OEC + 0.146 n_{\text{NHCOCONH},NH_2>1}$$
$$+ 0.124 P_{\alpha\text{-CH or } CH_3O-[C(NO_2)-CH-C(NO_2)]}, \tag{10.3}$$

where OEC, $n_{\text{NHCOCONH},NH_2>1}$ and $P_{\alpha\text{-CH or } CH_3O-[C(NO_2)-CH-C(NO_2)]}$ are the contributions of the elemental composition, the second and the third structural parameters, respectively. The value of $P_{\alpha\text{-CH or } CH_3O-[C(NO_2)-CH-C(NO_2)]}$ is 1.0 for the presence of either one α-CH (e. g. 2,4,6-trinitrotoluene) or methoxy group attached to one aromatic ring, or $CH_3O-[C(NO_2)-CH-C(NO_2)]$ group. The value of OEC is the optimized elemental composition of polynitro arenes with general formula $C_aH_bN_cO_d$, which can be obtained by $OEC = 2.74a - 1.48b - 2.31c - d$.

Example 10.3. 2,2′,2″,4,4′,6,6′,6″-Octanitro-1,1′:3′,1″-terphenyl (ONT) has the following molecular structure:

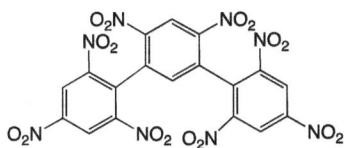

If equation (10.3) is used, the result is given as

$$OEC = 2.74a - 1.48b - 2.31c - d$$
$$= 2.74(18) - 1.48(6) - 2.31(8) - 16$$
$$= 5.96,$$

$$\log(E_a) = 2.25 + 0.0337OEC + 0.146n_{NHCOCONH,NH_2>1}$$
$$+ 0.124P_{\alpha\text{-CH or } CH_3O-[C(NO_2)-CH-C(NO_2)]}$$
$$= 2.25 + 0.0337(42.92) + 0.146(0) + 0.124(0)$$
$$= 2.451$$

$$E_a = 282.4 \text{ kJ/mol.}$$

The measured value of E_a is 281.58 kJ/mol [412].

10.1.4 Organic energetic compounds

For various organic energetic compounds, the results show that the important factors for predicting E_a can be grouped into additive and nonadditive structural parameters as follows [417]:

$$E_a = 166.36 + 2.85a - 21.2n_{OH} + 31.98E^+_{nonadd} - 44.93E^-_{nonadd}, \qquad (10.4)$$

where n_{OH} represents the number of hydroxyl groups; two functions E^+_{nonadd} and E^-_{nonadd} show the increasing and decreasing contribution of nonadditive structural parameters, respectively.

E^+_{nonadd}
The presence of some structural parameters can increase the activation energy and enhance the thermal stability of energetic compounds.
(1) Cyclic nitramines: The value of E^+_{nonadd} is 1.0 for cyclic nitramines which correspond to one of the following conditions:
   (a) ring is larger than a six-membered ring and contains two $-NNO_2$ groups;
   (b) ring larger than a four-membered ring;
   (c) bicyclic ring with only two $-NNO_2$ groups per ring.

These conditions are consistent with the correlation of activation energy in which the activation energies increase with increasing ring size [415].

(2) Nitroalkanes: For nitroalkanes containing molecular moieties $R–CH_2NO_2$ and $R–C(NO_2)_2–R'$, the values of E_{nonadd}^+ equal 2.0 and 1.0, respectively.

(3) Nitroaromatics:

(a) If the amino pyridine derivative, , triazine ring or four adjacent nitrogens in nitroaromatics are present, the value of E_{nonadd}^+ is 1.0;

(b) The value of E_{nonadd}^+ is 2.0 for the presence of the PNT——TNP molecular fragment where TNP is 2,46-trinitrophenyl.

E_{nonadd}^-

(1) Acyclic nitramines and cyclic nitramines containing small rings: For acyclic nitramines which contain only one $–NNO_2$ group in the form $Ar(or\ H)–N(NO_2)CH_3$, the value of E_{nonadd}^- is 0.75. The value of E_{nonadd}^- is equal to 0.4 for cyclic nitramines containing rings which are smaller than five membered rings, or five membered rings with more than one $N–NO_2$ group.

(2) Nitroaromatics:

(a) If the molecular moieties $–O(or\ S)–R(or\ Ar)$ are present, the value of E_{nonadd}^- equals 1.5;

(b) If the molecular fragment is present, the value of E_{nonadd}^- equals the number of this molecular moiety;

(c) The values of E_{nonadd}^- equal 1.5, 1.0 and 0.75 for the compound TNP–X where X is $–Cl$, $–NH–$ and $–N<$, respectively;

(d) For energetic compounds of the type TNP–Y–TNP where Y are $–N=N–$, $–CH_2–CH_2–$ and $–SO_2–$, the value of E_{nonadd}^- is 2.0.

(3) Presence of the nitrate group: if the $–ONO_2$ group is present, the value of E_{nonadd}^- is 0.3.

(4) The presence of the nitroso group: The value of E_{nonadd}^- is equal to 0.75.

Example 10.4. If equation (10.4) is applied to 2,4,6-trinitrobenzene-1,3,5-triol (TNPg) with the following molecular structure

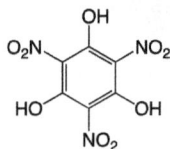

it gives

$$E_a = 166.36 + 2.85a - 21.2n_{OH} + 31.98E^+_{nonadd} - 44.93E^-_{nonadd}$$
$$= 166.36 + 2.85(6) - 21.2(3) + 31.98(0) - 44.93(0)$$
$$= 119.9 \, kJ/mol.$$

The measured value of E_a is 114.16 kJ/mol [412].

10.2 Heat of decomposition and temperature of thermal decomposition

Thermal analysis methods can be used to study the heat of decomposition and onset temperature [82, 206, 285, 418]. DSC [1, 419] is a typical example of an experimental screening test, which provides heats of decomposition with uncertainties in the measurement of about 5–10 % [420]. The exothermic onset temperature (the temperature at which the first deflection from the baseline is observed), thermal decomposition temperature and the temperature at which maximum mass loss occurs are three important parameters for assessing the heat sensitivity of different kinds of energetic compounds, and which give better reproducibility than the heats of decomposition. In the following sections, some methods for the prediction of these parameters are discussed.

10.2.1 Heat of decomposition of nitroaromatics

Some QSPR studies were undertaken to predict the heats of decomposition of nitroaromatic compounds. For 19 nitrobenzene derivatives, Saraf et al. [421] introduced a correlation based on the number of nitro groups (n_{NO_2}) with a fitting error of 8 %. Fayet et al. [422–424] used a series of preliminary multilinear models to derive some correlations from a data set of 22 molecules using some quantum chemical descriptors. Fayet et al. [425, 426] found very robust models by analyzing a more extended data set of 77 nitrobenzene derivatives. A suitable model was introduced for the whole diverse list of structures using a qualitative decision tree with high reliability [426]. For some nitrobenzenes which have no substituent in located in the *ortho* position to the nitro group, another complex model was introduced based on quantum mechanical calculations [426]. Three multilinear models were introduced for nitrobenzenes which have no substituents in the *ortho* position relative to the nitro group using a set of complex descriptors [427]. These methods are complex because they require special computer codes and expert users.

A simple correlation on the basis of n_{NO_2} has been introduced to calculate the heat of decomposition by considering inter- and intramolecular interactions rather using complex molecular descriptors as follows [428]:

$$- \Delta H_{\text{decom}} = -53.32 + 362 n_{NO_2} - 99.33 \Delta H^-_{\text{decom}} + 108.9 \Delta H^+_{\text{decom}}, \qquad (10.5)$$

where ΔH_{decom} is the heat of decomposition in kJ mol^{-1}; n_{NO_2} is the number of nitro groups; and $\Delta H^+_{\text{decom}}$ and $\Delta H^-_{\text{decom}}$ are two correcting functions. The presence of some molecular moieties can influence the values of the heat of decomposition and are referred to as $\Delta H^+_{\text{decom}}$ and $\Delta H^-_{\text{decom}}$, and described in the following sections.

$\Delta H^-_{\text{decom}}$

(1) Some molecular fragments in nitrobenzene:
   (a) Intramolecular hydrogen bonding: The value of $\Delta H^-_{\text{decom}}$ is 1.5 for the presence of only $-OH$ or $-CH_2-COOH$ groups *ortho* to the nitro group (or another hydrogen bonding group such as $-NH_2$).
   (b) Mono methyl derivatives: The value of $\Delta H^-_{\text{decom}}$ is 1.5. It was shown that the presence of intramolecular hydrogen bonding or a methyl group may decrease the heats of sublimation in nitroaromatic compounds [76, 98].
(2) The presence of halogens (or $-CF_3$) beside nitro groups in halogenated derivatives of nitrobenzene or polynitrobenzene: The value of $\Delta H^-_{\text{decom}}$ equals 3.0.
(3) The existence of $-COOH$ beside nitro groups in polynitrobenzene: The value of $\Delta H^-_{\text{decom}}$ is 4.0.

$\Delta H^+_{\text{decom}}$

(1) Some specific molecular moieties in nitrobenzene: if the $(Cl$ or $H)-C=O$, $-SCN$, $(NHNH_2)-C=O$ or $-C=C-$ groups are present, the values of $\Delta H^+_{\text{decom}}$ are 0.8, 2.0, 1.5 and 1.7, respectively.
(2) Some molecular fragments in polynitrobenzene: The values of $\Delta H^+_{\text{decom}}$ are 4.0, 3.0 and 2.0 for the presence of the $-C(=O)NH_2$, halogen and methyl groups, respectively.

Example 10.5. For 2-amino-4-nitrophenol with the following molecular structure

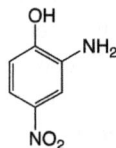

equation (10.5) gives

$$-\Delta H_{\text{decom}} = -53.32 + 362.0 n_{NO_2} - 99.33 \Delta H^-_{\text{decom}} + 108.9 \Delta H^+_{\text{decom}}$$

$$= -53.32 + 362.0(1) - 99.33(1.5) + 108.9(0)$$

$$= 159.6 \text{ kJ/mol}.$$

The measured value of $-\Delta H_{decom}$ is 130 kJ/mol [420]. The calculated value by complex QPRR model of Fayet et al. [426] is 238 kJ/mol (Dev = 108 kJ/mol).

10.2.2 Heats of decomposition of organic peroxides

Several QSPR methods containing complex descriptors were applied to explore:
(1) the relationship between the heat and temperature of decomposition of organic peroxides and their quantum mechanical descriptors [429, 430];
(2) the correlation between the self-accelerating decomposition temperature (SADT) of organic peroxides and their molecular structures [431] or their quantum mechanical properties [432]. These methods require specific computer codes and expert users.

It was found that molecular structures of organic peroxides can be used to predict the heat of decomposition of these compounds as follows [433]:

$$-\Delta H'_{decom} = 1551.45 - 41.13a + 1014.05P_{(HO-O...-O-OH)}$$
$$+ 640.13\alpha + 857.68\beta, \tag{10.6}$$

where $\Delta H'_{decom}$ is the heat of decomposition in $J\,g^{-1}$; $P_{(HO-O...-O-OH)}$ is 1.0 for the presence of two hydroperoxy functional groups in the molecular structure. The parameters α and β are nonadditive structural parameters, which can be specified as follows.

Definition of α: if aromatic rings, nonaromatic rings and nonaromatic rings containing methyl substituents are present, the values of α are 0.1, 0.2 and 0.6, respectively. If the organic peroxide contains the R fragment in its structure, the value of α equals 2.0.

Definition of β: if the $-O-\overset{O}{\underset{\|}{C}}-O-O-\overset{O}{\underset{\|}{C}}-O-$ molecular fragment is present in acyclic peroxides, β is 0.5. If two $(-O-O-R)$ or $-O-O-\overset{O}{\underset{\|}{C}}-R$ fragments are present, the value of β is 0.4 and 0.6, respectively.

Example 10.6. Di-tert-butyl peroxide has the following molecular structure.

The use of equation (10.6) gives

$$-\Delta H'_{decom} = 1551.45 - 41.13a + 1014.05P_{(HO-O...-O-OH)} + 640.13\alpha + 857.68\beta$$
$$= 1551.45 - 41.13(8) + 1014.05(0) + 640.13(0) + 857.68(0)$$
$$= 1222.4\,J/g.$$

There are two measured values of $-\Delta H'_{decom}$, i. e. 1082.5 [434] and 1175.0 J/g [432].

10.2.3 Onset temperature of polynitro arenes and organic peroxides as well as maximum loss temperature of organic azides

Since experimental data has been reported for the onset and maximum temperature of thermal decomposition for selected classes of energetic compounds, it has been possible to develop different methods to predict these values for other explosives. These methods are discussed in the following sections.

Onset temperature of polynitro arenes

For some subgroups of polynitro arenes, there are some relationships between the onset temperature and detonation characteristics [349, 400]. Using the DFT B3LYP/ 6-3-31G** method, it was found that there is a linear relationship between the onset temperature of thermal decomposition and the electronic charges of nitrogen atoms of selected subgroups of polynitro arenes [409]. For polynitro arenes, it has been shown that the following correlation can be used [435]:

$$T_{onset} = 571.17 + 30.63a - 21.29b + 32.57c - 43.11d$$
$$+ 15.98(n_{NH_2} - n_{NO_2}) + 50.69(|n_{TNB} - 2| - P_{TNB-CH_2-TNB}), \qquad (10.7)$$

where T_{onset} is the onset temperature in K; $(n_{NH_2} - n_{NO_2})$ is the difference between the number of amino and nitro groups in energetic compounds containing amino groups; $|n_{TNB} - 2|$ is the absolute value of the number of 1,3,5-trinitrobenzene rings minus two; P_{TNB-CH_2-TNB} is equal to one or zero for the presence or absence of $-CH_2-$ between two 1,3,5-trinitrobenzene aromatic rings, respectively.

Example 10.7. The use of equation (10.7) for the following energetic compound

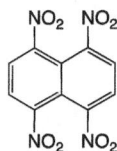

gives:

$$T_{onset} = 571.17 + 30.63a - 21.29b + 32.57c - 43.11d$$
$$+ 15.98(n_{NH_2} - n_{NO_2}) + 50.69(|n_{TNB} - 2| - P_{TNB-CH_2-TNB})$$
$$= 571.17 + 30.63(10) - 21.29(4) + 32.57(4) - 43.11(8)$$
$$+ 15.98(0) + 50.69(0)$$
$$= 577.7\ K.$$

The measured T_{onset} is 578.8 K [400].

Onset temperature of organic peroxides

For organic peroxides, it has been shown that the decomposition onset temperature can be expressed as a function of several structural parameters as [433]:

$$T_{onset} = 438.96 - 4.92d - 29.11P_{C=O} - 14.02\lambda_{sym} - 30.81\lambda', \tag{10.8}$$

where $P_{C=O}$ is 1.0 for the presence of the carbonyl group; and λ_{sym} is 1.0 for those peroxides that have the same fragments attached to each side of the $-O-O-$ bond, i. e. $R-O-O-R'$ where R=R'. The presence of some molecular fragments may also affect the values of T_{Dec}, which are incorporated as correcting factors. The parameter λ' represents the positive and negative contributions of various structural features which allows more reliable T_{onset} values to be obtained, and which can be defined based on the presence of the following molecular fragments:

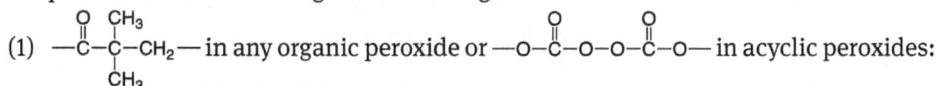

(1) in any organic peroxide or in acyclic peroxides:
The value of λ' is 1.0.

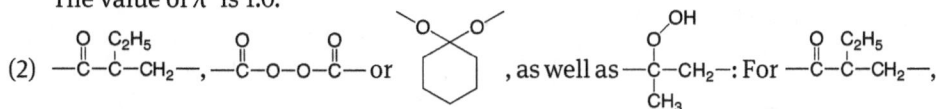

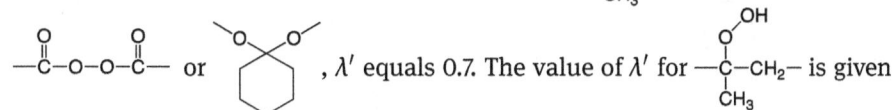

(2) , as well as For ,

, λ' equals 0.7. The value of λ' for is given

as: $\lambda' = 0.7 \times$ the number of , in molecular structure of peroxide.

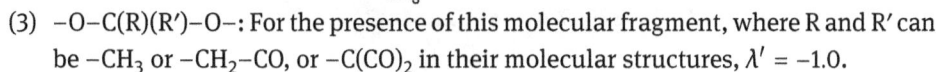

(3) $-O-C(R)(R')-O-$: For the presence of this molecular fragment, where R and R' can be $-CH_3$ or $-CH_2-CO$, or $-C(CO)_2$ in their molecular structures, $\lambda' = -1.0$.

Example 10.8. If equation (10.8) is used for the peroxide shown in Example 10.6, it gives

$$T_{onset} = 438.96 - 4.92d - 29.11\lambda_{C=O} - 14.02\lambda_{sym} - 30.81\lambda'$$
$$= 438.96 - 4.92(2) - 29.11(0) - 14.02(1) - 30.81(0)$$
$$= 415.1 \text{ K.}$$

There are two measured values of T_{onset}, i. e. 412.85 [434] and 426.15 K [432].

Temperature of maximum mass loss of organic azides

For different of organic azides, it was shown that the temperatures of maximum mass loss (T_{dmax}) can be given by [436]:

$$T_{dmax} = 405.57 + 1.3959b + 4.3222c + 33.670T^+_{dmax} - 32.515T^-_{dmax}, \qquad (10.9)$$

where T_{dmax} is in K; T^+_{dmax} and T^-_{dmax} are two correcting functions which are used to show the increasing and decreasing contribution of nonadditive structural parameters, respectively. The values of T^+_{dmax} and T^-_{dmax} are given in Tables 10.1 and 10.2 for different molecular fragments. As shown in Tables 10.1 and 10.2, the position of the azide group in an aromatic ring, neighboring groups and the presence of some specific functional groups are important parameters in predicting the values for T^+_{dmax} and T^-_{dmax}. Since the attachment of Cl–, –CH$_2$OH and –Ar *ortho* to –N$_3$ group may increase thermal stability, the contribution of T^+_{dmax} should be considered. The effect of T^+_{dmax} has also been considered for the presence of the –C(=O)O– group under certain conditions, and also for the attachment of the –N$_3$ group to tertiary carbon. The presence of the –CH$_2$–N(cyclic) molecular moiety, the existence of Cl *ortho* to the nitro group and the presence of –CO– *ortho* to the –N$_3$ group can decrease the thermal stability. Therefore, the presence of some molecular fragments or specific groups in an organic azide is responsible for increasing and decreasing the thermal stabilities.

Example 10.9. The use of equation (10.9) for 4-azido-3-nitro-2H-chromen-2-one with the following molecular structure

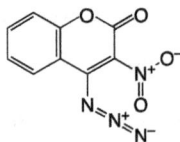

gives

$$T_{dmax} = 405.57 + 1.3959b + 4.3222c + 33.670T^+_{dmax} - 32.515T^-_{dmax}$$
$$= 405.57 + 1.3959(4) + 4.3222(4) + 33.670(1.0) - 32.515(0)$$
$$= 462.11 \text{ K.}$$

The measured value of T_{dmax} is 463.15 K [437].

Table 10.1: Summary of the contributions of T_{dmax}^+.

Molecular fragment	Condition	T_{dmax}^+	Example
(structure: 2-azido chromene fragment, H / N₃ / H or Cl)		0.7	2-azido-2H-chromen-2-one
(structure: azidophenyl fragment, N₃ / H or CH₃OH)		0.9	3-(3-azidophenyl)-6-chloropyridzin-4-ol
terrtiary carbon——N₃		1.0	3-azido-7-methyl-phenylquinoline-2,4-(1H,3H)-dione
(structure: N₃ / H / Ar fragment)		0.7	4-azido-2-chloro-3-phenylquinoline
(structure: O_2N, R, carbonyl/ester fragment)		1.0	4-azido-3-nitro-2H-chromen-2-one
O_2N——Ar——(carbonyl-O)		1.5	1,3-diazidopropan-2-yl 3,5dinitrobenzoate
O——(carbonyl)——(H₂C)n——Ar——(CH₂)n——(carbonyl)——O	$n \geq 0$	1.0	8-azidooctyl 2-(2-((8-azidooctyl)oxy)-2-oxoethyl)
(structure: diester fragment, R_1, R_2)		1.2	bis(1,3-diazidopropan-2-yl)-1H-indene-2,2(3H)-dicarboxylate)
N_3——CH_2——(carbonyl)——O		1.5	1,3-bis(azidoacetoxy)-2-azidoacetoxymethyl-2-ethylpropane

10.3 Deflagration temperature

The deflagration temperature is defined as the temperature at which a small sample of an energetic compound gets ignited [438]. This can be determined by heating 0.02 g of sample in a glass tube in a Wood's metal bath at a heating rate of 5 °C/min. Since various inter- and intramolecular parameters may affect the value of the deflagration temperature, a suitable correlation for estimating the deflagration temperature of energetic compounds containing $-NNO_2$, $-ONO_2$ or $-CNO_2$ groups was established as

Table 10.2: Summary of the contributions of T_{dmax}^-.

Molecular moieties	T_{dmax}^-	Example
N (cyclic) $\vert$ CH$_2$ $\vert$	1.0	7-azido-5-oxo-2,3-dihydro-1H,5H-pyrido[3,2,1-ij]quinoline-6-carbaldehyde
	1.0	4-azido-2-chloro-6-methyl-3-nitropyridine
	0.8	1-(2-azidophenyl)-1-ethanone

[48, 439]:

$$DT = 476.6 + 13.08a - 6.21d + 103.7F_{nonadd}^+ - 103.1F_{nonadd}^-, \qquad (10.10)$$

where DT is the deflagration temperature in K; two functions F_{nonadd}^+ and F_{nonadd}^- show the increasing and decreasing contribution of nonadditive structural parameters, respectively, which are specified in the following sections.

10.3.1 F_{nonadd}^+

(1) Cyclic nitramines containing rings which are larger than six-membered rings or carbocyclic nitroaromatics with

$$\frac{1}{3} \leq \frac{n_{NH_2}}{n_{NO_2}} \leq 1,$$

where n is the number of specified groups: The value of F_{nonadd}^+ is 1.0.
(2) Carbocyclic nitroaromatic compounds containing only alkyl substituents: The value of F_{nonadd}^+ is 0.5.
(3) The existence of other specific polar groups: The value of F_{nonadd}^+ equals 0.5 for the presence of a nitrate salt, –NHCONH– group, or cyclic energetic compounds containing >N–CO–N<, two >N–CO–N(NO$_2$)– or two –(O$_2$N)N–CO–N(NO$_2$)– groups. For the presence of a tertiary amine (or –O$^-$NH$_4^+$), the value of F_{nonadd}^+ equals 0.8.

10.3.2 F_{nonadd}^-

(1) The presence of azido, –N–OH, cyclic ether groups, as well as the presence of both >C(NO$_2$)$_2$ and –OH: For the presence of the –N$_3$, –N–OH or cyclic ether groups and

the presence of both of $>C(NO_2)_2$ and $-OH$ groups, the values of F_{nonadd}^- equal 0.7, 0.5, 0.3 and 1.0, respectively.

(2) Substituents containing $>NNO_2$ or $-NHNO_2$ groups attached to carbocyclic nitro-aromatic compounds: For the presence of $>NNO_2$ and $-NHNO_2$ groups, the values of F_{nonadd}^- equal 0.5 and 1.0, respectively.

10.3.3 Energetic compounds containing both F_{nonadd}^+ and F_{nonadd}^-

(1) Carbocyclic nitroaromatic compounds containing both nitrate and ether groups in their substituents: For the presence of one nitrate group, the value of F_{nonadd}^+ equals 0.5. The value of F_{nonadd}^- equals 0.5 for the presence of more than one nitrate group.

(2) Energetic compounds Ar–NH–Ar′: For secondary amines attached to two aromatic rings if Ar and Ar′ contain the N–O–N group as well as one nitro group or tetrazole ring, the value of F_{nonadd}^- is 1.0. Meanwhile, if Ar′ contains more than one nitro group, the value of F_{nonadd}^+ is 0.5.

Example 10.10. If equation (10.10) is used for azido-acetic-acid-3-(2-azido-acetoxy)-2-(2-azido-acetoxymethyl)-2-nitropropylester with the following molecular structure

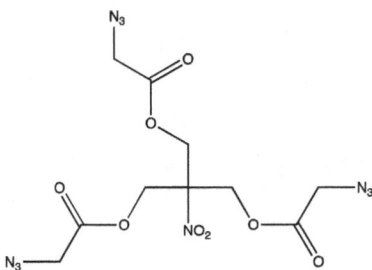

it gives

$$DT = 476.6 + 13.08a - 6.21d + 103.7F_{nonadd}^+ - 103.1F_{nonadd}^-$$
$$= 476.6 + 13.08(10) - 6.21(8) + 103.7(0) - 103.1(0.7)$$
$$= 488.5\,K.$$

The measured value of T_{dmax} is 487 K [440].

10.4 Thermal stability of selected classes of energetic ionic liquids and salts

Prediction of thermochemical parameters of a new energetic ionic liquid is of the utmost importance for their design to meet specific requirements and industrial applications. Several simple correlations have been developed for prediction activation energy and decomposition temperature (onset) of several classes of ionic liquids and salts including energetic derivatives, which are discussed here.

10.4.1 Predicting activation energy of thermolysis of some selected ionic liquids

A simple model has been developed for the prediction of activation energy of thermolysis of imidazolium, pyridinium, and phosphonium based ionic liquids, mainly based on TGA, through the structure of their anions and cations as follows [441]:

$$E_{a,IL} = 106.12 + 1.1574b_{cat} - 112.06i_{cat} - 6.6481e_{ani} + 23.036h_{ani} + 33.831i_{ani} + 47.681j_{ani}$$
$$- 58.688E_{a,IL}^- + 56.984E_{a,IL}^+ \tag{10.11}$$

where $E_{a,IL}$ is activation energy of thermolysis of the desired ionic liquid or salt in kJ/mol; b_{cat}, i_{cat}, and e_{cat} are the number of hydrogen, phosphorous, and fluorine atoms in cation, respectively; h_{ani}, i_{ani}, and j_{ani} are the number of sulfur, phosphorous, and bromine atoms in the anion, respectively; $E_{a,IL}^-$ and $E_{a,IL}^+$ are two correcting functions. Tables 10.3 and 10.4 show the values of $E_{a,IL}^-$ and $E_{a,IL}^+$ equal 1.0 only for some specific imidazolium-based ILS under certain conditions.

Table 10.3: The contribution of $E_{a,IL}^- = 1.0$ in ionic liquids or salts.

Cation	Anion
R' = Unsaturated alkyl group	Br⁻ HO₄S⁻, H₂O₄P⁻ Cl⁻

Table 10.4: The contribution of $E_{a,IL}^+ = 1.0$ in ionic liquids or salts.

Cation	Anion
R = Saturated alkyl group with carbon atoms less than 3	
R = Saturated alkyl group with carbon atoms greater than 5	Br$^-$

Example 10.11. The use of equation (10.11) for the following energetic ionic liquid

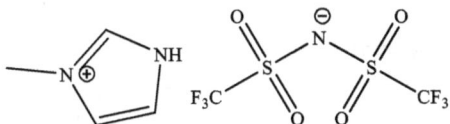

gives:

$$E_{a,IL} = 106.12 + 1.1574b_{cat} - 112.06i_{cat} - 6.6481e_{ani} + 23.036h_{ani} + 33.831i_{ani} + 47.681j_{ani}$$
$$- 58.688E_{a,IL}^- + 56.984E_{a,IL}^+$$
$$= 106.12 + 1.1574 \times 7 - 112.06 \times 0 - 6.6481 \times 6 + 23.036 \times 2 + 33.831 \times 0$$
$$+ 47.681 \times 0 - 58.688 \times 0 + 56.984 \times 0 = 120.4 \, \text{kJ/mol}$$

The measured value of $E_{a,IL}$ is 102.7 kJ/mol [442].

10.4.2 Decomposition temperature of imidazolium-based energetic ionic liquids or salts

The temperature of decomposition of imidazolium based energetic ionic liquids or salts can be correlated with their molecular structures as follows [443]:

$$T_{onset,IL} = \frac{1276b_{cat} - 1875d_{cat} + 4439e_{cat}}{Mw_{cat}}$$
$$+ \frac{4542a_{ani} + 5255c_{ani} + 4958d_{ani} + 8244e_{ani} + 20793f_{ani} + 30766h_{ani}}{Mw_{ani}}$$

$$+ \frac{6313 i_{ani} + 6811 l_{ani}}{Mw_{ani}}$$

$$+ 116.0 T^+_{onset,IL} - 126.2 T^-_{onset,IL} \qquad (10.12)$$

where $T_{onset,IL}$ is decomposition temperature in K; b_{cat}, d_{cat} and e_{cat} are the number of hydrogen, oxygen, and fluorine atoms in imidazolium cation derivatives, respectively; a_{ani}, c_{ani}, d_{ani}, e_{ani}, f_{ani}, h_{ani}, i_{ani} and l_{ani} are the number of carbon, nitrogen, oxygen, fluorine, chlorine, sulfur, phosphorus and boron atoms in anion part of imidazolium ionic liquids, respectively; Mw_{cat} and Mw_{ani} are the molecular weights of cation and anion, respectively; $T^+_{onset,IL}$ and $T^-_{onset,IL}$ are correcting functions that can adjust large negative and positive deviations of the first two ratios of equation (10.12) from the measured data. Tables 10.5 and 10.6 show the values of $T^+_{onset,IL}$ and $T^-_{onset,IL}$ for the presence of specific cations and anions in some imidazolium-based ionic liquids or salts.

Example 10.12. The use of equation (10.12) for 3-(but-2-ynyl)-1-methyl-1H-imidazol-3-ium azide with the following structure

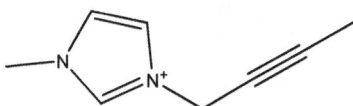

Chemical Formula: $C_8H_{11}N_2{}^+$
Molecular Weight: 135.19

Chemical Formula: $N_3{}^-$
Molecular Weight: 42.02

gives:

$$T_{onset,IL} = \frac{1276 b_{cat} - 1875 d_{cat} + 4439 e_{cat}}{Mw_{cat}}$$

$$+ \frac{4542 a_{ani} + 5255 c_{ani} + 4958 d_{ani} + 8244 e_{ani} + 20793 f_{ani} + 30766 h_{ani}}{Mw_{ani}}$$

$$+ \frac{6313 i_{ani} + 6811 l_{ani}}{Mw_{ani}} + 116.0 T^+_{onset,IL} - 126.2 T^-_{onset,IL}$$

$$= \frac{1276 \times 11 - 1875 \times 0 + 4439 \times 0}{135.19}$$

$$+ \frac{4542 \times 0 + 5255 \times 3 + 4958 \times 0 + 8244 \times 0 + 20793 \times 0 + 30766 \times 0}{42.02}$$

$$+ \frac{6313 \times 0 + 6811 \times 0}{42.02} + 116.0 \times 0 - 126.2 \times 0.7 = 391 \, K$$

The measured value of $T_{onset,IL}$ is 388 K [444].

Table 10.5: Specific cation/anion moieties for estimation of $T^+_{onset,IL}$.

Cation	Anion	$T^+_{onset,IL}$
R_1, R_2 and R_3 may contain a fluorine atom		0.9
		0.9
		0.9
		1.0
R is the saturated alkyl group		0.5
$n<3$		0.6
		—
		0.7

Table 10.6: Specific cation/anion moieties for estimation of $T_{onset,IL}^{-}$.

Cation	Anion	$T_{onset,IL}^{-}$
R containing a double bond		0.4
R containing a triple bond		0.7
		0.5
R contains a triple bond		0.5
		0.9
R_1 and R_2 contain double bounds	NO_3^{-}	0.6
, R_1 contains fluorine atoms		0.5

10.5 Decomposition temperature of azole-based energetic compounds

A reliable correlation has been introduced for the prediction of thermal decomposition temperature of azole-based energetic compounds by their structural parameters as follows [445]:

$$T_{\text{onset,azole}} = 547.75 + \frac{7777.96a - 2138.50c}{Mw} - 52.40\frac{d}{a} - 118.75\frac{b}{c} - 54.36n_{N_3}$$

$$- 60.85n_{\text{NH-NO}_2} + 102.62n_{\text{C=NH}} + 51.83T^+_{\text{onset,azole}}$$

$$- 65.85T^-_{\text{onset,azole}} \tag{10.13}$$

where $T_{\text{onset,azole}}$ is the decomposition temperature of the desired azole-based energetic compound in K; a, b, c, and d are the number of carbon, hydrogen, nitrogen and oxygen atoms, respectively; Mw is the molecular weight of the compound; n_{N_3}, $n_{\text{NH-NO}_2}$, and $n_{\text{C=NH}}$ are the number azido, nitroamino, and imino groups, respectively; $T^+_{\text{onset,azole}}$ and $T^-_{\text{onset,azole}}$ are nonadditive structural parameters which have been specified in Tables 10.7 and 10.8.

Example 10.13. The use of equation (10.13) for 1-(2-(5-amino-3-nitro-1H-1,2,4-triazol-1-yl)ethyl)-1H-tetrazol-5-amine with the following structure

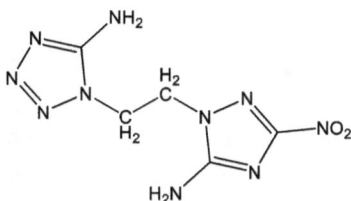

Chemical Formula: $C_5H_8N_{10}O_2$
Molecular Weight: 240.19

gives:

$$T_{\text{onset,azole}} = 547.75 + \frac{7777.96a - 2138.50c}{Mw} - 52.40\frac{d}{a} - 118.75\frac{b}{c} - 54.36n_{N_3}$$

$$- 60.85n_{\text{NH-NO}_2} + 102.62n_{\text{C=NH}} + 51.83T^+_{\text{onset,azole}} - 65.85T^-_{\text{onset,azole}}$$

$$= 547.75 + \frac{7777.96 \times 5 - 2138.50 \times 10}{240.19} - 52.40\frac{2}{5} - 118.75\frac{8}{10} - 54.36 \times 0$$

$$- 60.85 \times 0 + 102.62 \times 0 + 51.83 \times 0 - 65.85 \times 0 = 504.7 \text{ K}$$

The measured value of $T_{\text{onset,azole}}$ is 550 K [446].

Table 10.7: Specific structural moieties for estimation of $T^+_{onset,azole}$.

Structural moieties	Condition	$T^+_{onset,azole}$
(triazole structures with R_1, R_2, O_2N)	R_1 and $R_2 = NH_2$ $R_1 = H$, $R_2 = NHNO_2$ $R_1 = NH_2$, $R_2 = NO_2$	1.0
(tetrazine ring)	The presence of tetrazine in the polyciles compounds	1.0
(bis-pyrazole with R_1, R_2, R_3, NO_2)	R_1, R_2, and $R_3 = NO_2$ or N_3 R_1 and $R_2 = NO_2$ or N_3 $R_3 = Cl$	0.70 2.0
(fused bicyclic pyrazole with R_1, R_2)	$R_1 = CH_3N_3$ $R_2 = NO_2$	0.70
(ethylene-linked bis-pyrazole with R_1, R_2, R_3)	R_1, R_2, and $R_3 = NO_2$ or NH_2 or Cl	0.90
(R_1–O–R_2 ether)	R_1 and R_2 are nitro pyrazole or nitro triazole	1.0
(pyrazole with NO_2, O_2N, R_1, R_2)	R2 = H or NO_2 or NH_2 R1 = H or NH_2 or CH_3	1.0
(triazole with NO_2, positions 1 and 2)	Situation 1 = –H or –CH_2–N(NO_2)–CH_3 Situation 2 = –NH_2 or C(NO_2)$_3$ or –CO	1.0

10.6 Summary

This chapter introduced different methods for predicting: the activation energies of low-temperature non-autocatalyzed thermolysis, heats of decomposition, exothermic

Table 10.8: Specific structural moieties for estimation of $T^-_{onset,azole}$.

Structural moieties	Condition	$T^-_{onset,azole}$
R_2, R_3, R_3, R_1 (pyrazole ring)	Repeating of this moiety in the R_1 situation with a C–C bridge or the presence –OH substitution in the R_1 situation, i. e.: $R_1 =$ (structure with R_2, R_3, R_4) or $R_1 = -OH$; R_2 and $R_3 = -NO_2$ or $-N_3$	0.50
(tetrazole ring) NHCH$_2$C(NO$_2$)$_3$	The presence of two or three of this fragment in a molecule which is separated from each other by more than two atoms	0.50
(tetrazole ring) NNO$_2$	1. In the two-cycle molecules, the presence of one nitroiminotetrazole 2. Two nitroiminotetrazole cycles separated by at least four carbon atoms	0.90
	In the one cycle compounds, the presence nitroimino substitution	0.50
	In the one cycle compounds, the presence nitroimino, and cyclopentane substitutions	1.5
$R_1-N=N-R_2$	R_1 and R_2 = triazole or tetrazole ring	1.0
	R_1 and R_2 = tetrazole ring	2.0
(pyrazole ring) NO$_2$, R_2, O$_2$N, R_1	$R_1 = -H$ or $-NH_2$ or $-CH_3$ $R_2 = -H$ or $-N_3$	1.0
(triazole ring) R_2, R_1, R_3	$R_1 = -H$ or $-NH_2$ $R_2 = -H$ or $-NO_2$ or $-CH(NO_2)_2$	1.0
(furan ring)	The presence of $-CO$, $-CH_2-ONO_2$, and furan fragment in a compound, simultaneously	1.0
	The presence of $-CH_2-ONO_2$ and triazole ring in a compound, simultaneously	0.50

onset temperatures, thermal decomposition temperatures of polynitro arenes and the temperature at which the maximum loss of mass occurs, as well as the deflagration temperature. Equations (10.1), (10.2), (10.3) and (10.4) were used to predict the activation energies of low-temperature, non-autocatalyzed thermolysis of nitroparaffins, nitramines and organic energetic compounds, respectively. Equation (10.4) is a general correlation, which can be applied to a wide range of energetic compounds including those energetic compounds that satisfy equations (10.1) to (10.3), but it is more complex. Equations (10.5) and (10.6) are used for the prediction of the decomposition temperatures of polynitro arenes and organic peroxides, respectively. Equations (10.7) and (10.8) can be applied for prediction of onset temperature of polynitro arenes and organic peroxides, respectively. Equation (10.9) is used to estimate the temperatures at which maximum mass loss occurs for organic azides. Equation (10.10) is a simple, reliable correlation for the prediction of the deflagration temperature of organic energetic compounds containing the $-NNO_2$, $-ONO_2$ or $-CNO_2$ functional groups. Equation (10.11) provides a simple method for calculation of activation energy of thermolysis of imidazolium, pyridinium, and phosphonium based ionic liquids. Equation (10.12) gives the temperature of decomposition of imidazolium based energetic ionic liquids or salts. Equation (10.13) introduces a reliable correlation for prediction of thermal decomposition temperature of azole based energetic compounds.

11 Relationships between different sensitivities

In previous chapters, different methods for the prediction of the impact, shock, electric spark, friction and heat sensitivities have been discussed. Among different sensitivity tests, the impact sensitivity test is extremely easy to implement. Since the results of the impact sensitivity test greatly depend on the conditions under which the tests are performed, the experimental data from impact sensitivity tests is not often reproducible. Hot spots in an energetic compound may contribute to initiation in the drop weight impact test. Due to the large number of impact sensitivity data which has been reported, some efforts have been undertaken to correlate the impact sensitivity of selected classes of energetic compounds to the other sensitivities [292, 359, 447–451]. Wu and Huang [452] used a complex micro-mechanics model to describe hot spot formation in the energetic crystal powders of the two well-known explosives HMX and PETN subjected to drop-weight impact. Many attempts have been undertaken to illustrate the mechanism of initiation an energetic material by impact stimulus, but this feature is not yet fully understood. In this chapter, several simple relationships between different sensitivities are demonstrated.

11.1 Relationship between impact sensitivity of energetic compounds and activation energies of thermal decomposition

It was indicated that initiation of the decomposition of an energetic compound containing $-NNO_2$, $-ONO_2$ and $R-NO_2$ (or $Ar-NO_2$) groups through impact and heat stimuli can be related as follows [292]:

$$E_{IS} = 18.07 - 0.1130E_a + 14.68(b/d) + 22.65E_{IS}^{++} - 11.30E_{IS}^{--}, \tag{11.1}$$

where E_a is the activation energy of low-temperature non-autocatalyzed thermolysis in kJ mol^{-1}; E_{IS} is the impact sensitivity in J; two correcting functions E_{IS}^{++} and E_{IS}^{--} show the increasing and decreasing contribution of nonadditive structural parameters, re-

https://doi.org/10.1515/9783110740158-011

spectively. The presence of some molecular moieties can increase or decrease the impact sensitivity values of different classes of energetic compounds through E_{IS}^{++} and E_{IS}^{--}, which are described in the following sections.

11.1.1 Nitroaromatics

(1) $-NH_2$ group: If per aromatic ring, then the values of E_{IS}^{++} are 0.9, 3.2 and 4.8 for n_{NH_2} = 1, 2 and 3 per aromatic ring, respectively.
(2) One 2,4,6-trinitrophenyl (TNP) in form TNP–X: The values of E_{IS}^{++} equal 0.5, 1.0, 1.5 for X = –NH–R (or –OH), –R and –OR, respectively.
(3) Two aromatic rings in form Ar–X–Ar or Ar–Ar: The value of E_{IS}^{--} equals 0.5 for X = –S– and –CH$_2$–R– except for 2,2′,4,4′,6,6′-hexanitro-1,1′-biphenyl (HNB).
(4) Polynitrobenzene containing more than one alkyl group: The value of E_{IS}^{--} is 1.0.
(5) Polynitronaphthalene: The value of E_{IS}^{++} is 0.5.

11.1.2 Nitramines

(1) Cyclic nitramines: For rings containing up to six ring atoms where , the value of E_{IS}^{++} is 0.5. For cyclic nitramines which have rings larger than six mebered ring and in which

$$0.75 \le \frac{n_{NNO_2}}{n_{CH_2}} < 1 \quad \text{and} \quad \frac{n_{NNO_2}}{n_{CH_2}} < 0.75,$$

the values of E_{IS}^{--} equal 1.0 and 2.0, respectively. The value of E_{IS}^{--} is equal to 0.5 for cyclic nitramines which are smaller than six membered rings and in which

$$\frac{n_{NNO_2}}{n_{CH_2}} \ge 1.$$

(2) Acyclic nitramines: For acyclic nitramines in which the number of $-N(NO_2)-CH_2-N(NO_2)-$ groups > 1 and ≤ 1, the values of E_{IS}^{--} and E_{IS}^{++} equal 0.75 and 1.75, respectively. The value of E_{IS}^{--} equals 1.0 for the presence of the $-N(NO_2)-CH_2-N(NO_2)-$ molecular fragment.

Example 11.1. 1,5-Endomethylene-3,7-dinitro-1,3,5-tetraazacyclooctane (DPT) has the following molecular structure:

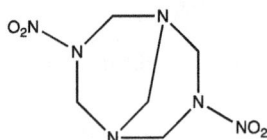

The measured E_a is 192.3 kJ/mol [412]. The use of equation (11.1) gives

$$E_{IS} = 18.07 - 0.1130E_a + 14.68(b/d) + 22.65E_{IS}^{++} - 11.30E_{IS}^{--}$$
$$= 18.07 - 0.1130(192.3) + 14.68(10/4) + 22.65(0.5) - 11.30(0)$$
$$= 10.44\,J.$$

The measured E_{IS} is 10.20 J [447].

11.2 Relationship between electric spark sensitivity and impact sensitivity of nitroaromatics

For nitroaromatics, it was shown that the electric spark and impact sensitivities can be correlated as [359]:

$$E_{ES} = 6.17 + 0.0797E_{IS} + 10.1E_{cor}^+ - 3.21E_{cor}^-, \tag{11.2}$$

where E_{ES} is the electric spark sensitivity in J; E_{cor}^+ and E_{cor}^- are two correcting functions that have been used to adjust large deviations of E_{ES} and E_I, and which can be specified as follows.

E_{cor}^+: The values of E_{cor}^+ are 1.8, 0.5 and 0.9 for the presence of the −OR group, two − OH groups and the attachment of more than one nitroaromatic ring to another nitroarmatic ring, respectively.

E_{cor}^-: The value of E_{cor}^- equals 1.0 for the attachment of only one CH_x− or Ar− to the aromatic ring in the case of CHNO nitroaromatics.

Example 11.2. If equations (6.11) and (6.12) are used to predict the impact sensitivity of 2-methyl-4-[(3-methyl-2,4,6-trinitrophenyl)thio]-1,3,5-trinitrobenzene (DIMEDIPS) with the following molecular structure

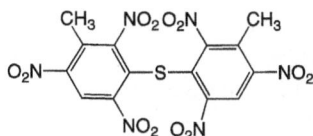

$E_{IS} = 26.05$ J is obtained. The use of this value in equation (11.2) gives

$$E_{ES} = 6.17 + 0.0797E_{IS} + 10.1E_{cor}^+ - 3.21E_{cor}^-$$
$$= 6.17 + 0.0797(26.05) + 10.1(0) - 3.21(0)$$
$$= 8.25\,J.$$

The measured E_{ES} is 8.57 J [348].

11.3 A general correlation between electric spark sensitivity and impact sensitivity of nitroaromatics and nitramines

It was shown that electric spark sensitivity based on RDAD instrument can be correlated with the impact sensitivity of nitramines [453]. Examination of various molecular moieties of nitroaromatics and nitramines has shown that the following general equation can be used to correlate their electric spark sensitivity based on RDAD instrument and impact sensitivity as [454]:

$$E_{ES} = 6.16 + 0.0843E_{IS} + 8.16E^+_{ES,NAr,NiA} - 3.43E^-_{ES,NAr,NiA} + 9.48E_{N\ excess}, \qquad (11.3)$$

where $E^+_{ES,NAr,NiA}$, $E^-_{ES,NAr,NiA}$ and $E_{N\ excess}$ are correcting functions. The values of $E^+_{ES,NAr,NiA}$ and $E^-_{ES,NAr,NiA}$ are given in Table 11.1.

Table 11.1: Contribution of $E^+_{ES,NAr,NiA}$ and $E^-_{ES,NAr,NiA}$.

Organic compounds containing energetic groups	$E^+_{ES,NAr,NiA}$	$E^-_{ES,NAr,NiA}$	Condition
Cyclic nitramine	0	1.0	Equal numbers of >NNO$_2$ and >CH$_2$ or $\diagdown$CH groups in form −NNO$_2$−CHx−NNO$_2$− with more than five-membered rings
	0.8	0	Equal numbers of >NNO$_2$ and −CH$_2$−CH$_2$−
	0.6	0	The presence of −O− group
Acyclic nitramine	0.8	0	The presence of −C(=O)−O− group
Nitroaromatics	0	1.0	(a) Direct attachment of two nitroaromatic rings or the presence of specific group (X) between nitroaromatic ring (Ar) where X is >CH2, >NH, −CH$_2$=CH$_2$− and −CH$_2$−CH$_2$− (b) The attachment both of alkyl and hydroxyl groups to 2,4,6-trinitrobenzene
	2.2	0	The attachment of −OR group to carbocyclic nitroaromatics
	0.6	0	The attachment of two −OH groups to one nitroaromatic ring
	1.2	0	Direct attachment of three nitroaromatic rings
	1.0	0	The presence of −NH−C(=O)−C(=O)−NH− The attachment of −OR group ortho to nitrogen atom ring
	0.5	0	The presence of −SO$_2$− group

The value of $E_{N\ excess}$ is for presence of excess nitrogen atoms for cyclic nitramines containing excess nitrogen atoms besides −NNO$_2$ and −NO$_2$.

Example 11.3. The reported E_{IS} of 1,4-dinitrotetrahydroimidazo[4,5-d]imidazole-2,5(1H,3H)-dione (DINGU) with the following molecular structure is 5.55 J [330].

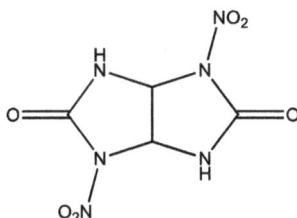

The use of this value in equation (11.3) gives:

$$E_{ES} = 6.16 + 0.0843E_{IS} + 8.16E^+_{ES,NAr,NiA} - 3.43E^-_{ES,NAr,NiA} + 9.48E_{N\,excess}$$
$$= 6.16 + 0.0843(5.55) + 8.16(0) - 3.43(0) + 9.48(1) = 16.11J$$

The measured E_{ES} is 15.19 J [364].

11.4 Relationship between electric spark sensitivity and activation energy of the thermal decomposition of nitramines

For various nitramines, it has been shown that the mechanism of spark energy transfer can be related to the activation energy of low-temperature nonautocatalyzed thermolysis as [450]:

$$E_{ES} = 9.826 - 0.047E_a + 7.432(a/d) + 7.680E^+_{ES,corr}, \tag{11.4}$$

where $E^+_{ES,corr}$ is an increasing factor. The value of $E^+_{ES,corr}$ equals 1.0 for the presence of the following molecular fragment or condition:
(1) cyclic –O–, >N– and –C(=O)–N–;
(2) acyclic –C(=O)–O–;
(3) acyclic nitramines with more than one N–NO$_2$ functional group in which

$$\frac{n_{NNO_2}}{n_{CH_2}} \geq 2;$$

(4) for cyclic nitramines with more than one N–NO$_2$ group in which

$$\frac{n_{NNO_2}}{n_{NCH_2CH_2N}} \leq 1,$$

where $n_{NCH_2CH_2N}$ is the number of –NCH$_2$CH$_2$N– groups.

Example 11.4. The molecular structure of 2,4,6-trinitro-2,4,6-triazaheptane (ORDX) is given as

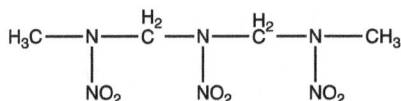

$$H_3C-N-\underset{\underset{NO_2}{|}}{\overset{H_2}{\overset{|}{C}}}-N-\underset{\underset{NO_2}{|}}{\overset{H_2}{\overset{|}{C^2}}}-N-CH_3$$
$$\underset{NO_2}{|}$$

If the measured E_a of ORDX is 179.2 kJ/mol [407], the use of equation (11.4) gives

$$E_{ES} = 9.826 - 0.047E_a + 7.432(a/d) + 7.680E^+_{ES,corr}$$
$$= 9.826 - 0.047(179.2) + 7.432(4/6) + 7.680$$
$$= 6.36 \text{ J.}$$

The measured E_{ES} is 8.08 J [364].

11.5 Correlation of the electrostatic sensitivity and activation energies for the thermal decomposition of nitroaromatics

It was shown that the following equation can be introduced as a suitable correlation for predicting the relationship between the electric spark sensitivity and activation energy of low-temperature non-autocatalyzed thermolysis [448]:

$$E_{ES} = -9.72 + 0.034E_a + 7.41\left(\frac{c+d}{a}\right) + 3.49E^+_{ES,Ar} - 2.50E^-_{ES,Ar}, \qquad (11.5)$$

where the functions $E^+_{ES,Ar}$ and $E^-_{ES,Ar}$ show increasing and decreasing contribution of structural parameters, respectively, which depend on the attachment of some groups to TNP which increase and decrease the electrostatic sensitivity. They are defined as follows.

(1) One TNP: For compounds with general structure TNP(X)$_2$ or TNP(X)$_3$, where X is $-CH_3$, $-NH_2$ or $-OH$, the values of $E^+_{ES,Ar}$ equal 1.75, 0.75 and 1.25, respectively. If $-Cl$ is present, the value of $E^-_{ES,Ar}$ equals 2.0.

(2) Two TNP: For energetic compounds with general formula TNP–Y–TNP, where Y is $-SO_2-$, $-HN-\overset{\overset{O}{\|}}{C}-\overset{\overset{O}{\|}}{C}-NH-$ or TNP, the value of $E^+_{ES,Ar}$ equals 1.40. If the $-NH-$ or $-S-$ groups are present, or if two TNP groups are directly attached together, the value of $E^-_{ES,Ar}$ equals 1.25.

Example 11.5. 1-Methyl-3-hydroxy-2,4,6-trinitrobenzene (TNCr) has the following molecular structure:

The structure shows a benzene ring with substituents: CH₃ at top, O₂N (left), NO₂ (right), OH (lower right), NO₂ (bottom).

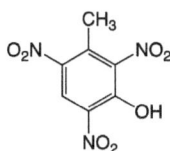

The measured E_a is 192.46 kJ/mol [364]. The use of equation (11.5) gives

$$E_{ES} = -9.72 + 0.034E_a + 7.41\left(\frac{c+d}{a}\right) + 3.49E_{ES,Ar}^+ - 2.50E_{ES,Ar}^-$$

$$= -9.72 + 0.034(192.46) + 7.41\left(\frac{3+7}{7}\right) + 3.49(0) - 2.50(0)$$

$$= 7.41\,J.$$

The measured E_{ES} is 5.21 J [364].

11.6 Relationship between the activation energy of thermolysis and friction sensitivity of cyclic and acyclic nitramines

The elemental composition and some structural parameters can be used to correlate friction sensitivity with the activation energy of thermolysis for cyclic and acyclic nitramines as [449]

$$FS = 212.0 + 32.67a - 10.21c - 14.50d - 85.07E_a/Mw$$
$$+ 81.92FS^+ - 48.19FS^-, \tag{11.6}$$

where functions FS^+ and FS^- show the presence of some structural parameters that can increase and decrease the predicted friction sensitivity values on the basis of the elemental composition and E_a/Mw. For cyclic and cage nitramines as well as acyclic nitramines, the correcting functions FS^+ and FS^- are defined as follows.
(1) Cyclic and cage nitramines: If the ratio of the number of $-NNO_2$ groups to the number of $-CH_2-$ groups, i. e. n_{NNO_2}/n_{CH_2}, in cyclic nitramines equals 1.0, the value of FS^+ is 0.8. The value of FS^- equals 0.7 for those compounds in which the ratios of n_{NNO_2}/n_{CH_2} in cyclic nitramines and n_{NNO_2}/n_{CH} in cage or unsaturated nitramines are less than 1.0, where n_{CH} is the number of $>CH-$ or $=CH-$ groups.
(2) Acyclic nitramines:
    (a) For compounds with general formula $TNP-N(NO_2)-R$, $FS^+ = 1.4$.
    (b) For compounds containing the neutral tetrazole molecular moieties, the values of FS^- and FS^+ are equal to 0.8 and 1.0 for the absence and presence of the $R-N<$ group attached to the tetrazole ring, respectively.
    (c) For compounds with general formula $-RN-(CH_2)_n-NR-$, $FS^- = 1.4$.

Example 11.6. 1-Methyl-5-nitriminotetrazole has the following molecular structure:

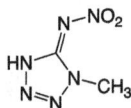

The calculated E_a is 189.16 kJ/mol [417]. The use of equation (11.6) gives

$$
\begin{aligned}
FS &= 212.0 + 32.67a - 10.21c - 14.50d - 85.07E_a/Mw \\
&\quad + 81.92FS^+ - 48.19FS^- \\
&= 212.0 + 32.67(2) - 10.21(6) - 14.50(2) - 85.07(189.16/144.09) \\
&\quad + 81.92(1.0) - 48.19(0) \\
&= 157.3\,\text{N}.
\end{aligned}
$$

The measured FS using the BAM friction tester is 160 N [455].

11.7 Relationship between shock sensitivity of nitramine energetic compounds based on small-scale gap test and their electric spark sensitivity

An attempt has been done to correlate shock sensitivity to electric spark sensitivity based on the RDAD instrument of nitramine compounds [456]. Tan et al. [457] applied an Mn–Cu manometer to measure the output pressures of shock waves. They improved a set of small-scale gap tests to pass through aluminum gaps with different thicknesses for drawing a standard curve. For the tested explosive pillars, they measured the thickness of aluminum gaps with the calibrated set in terms of "go" or "no go". Since the reported data of Tan et al. [457] and Storm et al. [336] are different from each other, there is a linear correlation between $P_{90\%\,\text{TMD}}$, which is calculated by equation (8.1), and the improved method of Tan et al. [457] as follows [458]:

$$
(x\,Gap)_{90\%\,\text{TMD}} = 12.50 - 0.1220P_{90\%\,\text{TMD}}, \tag{11.7}
$$

where $(x\,Gap)_{90\%\,\text{TMD}}$ is the thickness of aluminum in mm at 90 % of theoretical maximum density (TMD). Thus, equation (11.7) can convert the calculated shock sensitivity based on NSWC data using Navy small-scale gap test method of Storm et al. [336] to the improved method of Tan et al. [457].

It was shown that shock sensitivities at 90 %, 95 %, and 98 % of TMD and electric spark sensitivities based on the measurements of RDAD can be correlated for nitroaromatics and nitramines as:

$$
P_{90\%\,\text{TMD}} = 8.62 + 0.4970E_{ES} \tag{11.8}
$$

$$P_{95\% \text{ TMD}} = 12.29 + 0.7583E_{ES} - 9.113C_{NH_2} \tag{11.9}$$

$$P_{98\% \text{ TMD}} = 15.43 + 0.9723E_{ES} - 16.17C_{NH_2}, \tag{11.10}$$

where $P_{90\% \text{ TMD}}$, $P_{95\% \text{ TMD}}$, and $P_{98\% \text{ TMD}}$ are the pressures in kbar required to initiate material pressed to 90 %, 95 %, and 98 % of TMD, respectively; C_{NH_2} shows the contribution of amino groups in nitroaromatics. The values of C_{NH_2} are 1.0, 2.0, and −2.0 for the presence of one, two and three amino groups, respectively.

Example 11.7. 1,3,5-Trinitro-1,3,5-triazinane (RDX) has the following molecular structure:

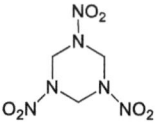

The use of equation (8.1) gives:

$$P_{90\% \text{ TMD}} = 16.79 + 2.262(a + b/2 - d) - 6.314E^0_{\alpha CH/NNO_2} + 17.72(1.93n_{NH_2} - n_{NO_2})_{pure}$$

$$= 16.79 + 2.262(3 + 6/2 - 6) - 6.314(1) + 17.72(0)$$

$$= 10.48 \text{ kbar}$$

The use of equation (11.7) gives:

$$(x \, Gap)_{90\% \text{ TMD}} = 12.50 - 0.1220P_{90\% \text{ TMD}} = 12.50 - 0.1220(10.47) = 11.22 \text{ mm}$$

The measured $(x \, Gap)_{90\% \text{ TMD}}$ is 11.68 mm [457].

The use of equation (7.3) provides:

$$E_{ES} = 3.460 + 6.504(a/d) - 4.059C_{CH_2NNO_2>2,C(= O)(O \text{ or } NH)}$$

$$= 3.460 + 6.504(3/6) - 4.059(1)$$

$$= 2.65 \, J$$

The use of equations (11.8), (11.9), and (11.10) gives:

$$P_{90\% \text{ TMD}} = 8.62 + 0.4970E_{ES} = 8.62 + 0.4970(2.65) = 9.94 \text{ kbar}$$

$$P_{95\% \text{ TMD}} = 12.29 + 0.7583E_{ES} - 9.113C_{NH_2} = 12.29 + 0.7583(2.65) - 9.113(0)$$

$$= 14.30 \text{ kbar}$$

$$P_{98\% \text{ TMD}} = 15.43 + 0.9723E_{ES} - 16.17C_{NH_2} = 15.43 + 0.9723(2.65) - 16.17(0)$$

$$= 18.01 \text{ kbar}$$

The reported values of $P_{90\% \text{ TMD}}$, $P_{95\% \text{ TMD}}$, and $P_{98\% \text{ TMD}}$ for RDX are 10.97, 15.77, and 20.35 kbar, respectively [336].

11.8 Summary

The relationships between the impact, electric spark, friction sensitivities and activation energies of low-temperature non-autocatalyzed thermolysis were discussed in this chapter. Equation (11.1) correlates the impact sensitivity of an energetic compound containing $-NNO_2$, $-ONO_2$ or $R-NO_2$ (or $Ar-NO_2$) groups with its activation energy of low-temperature non-autocatalyzed thermolysis. Equation (11.2) gives another correlation between the electric spark sensitivity based on RDAD instrument and impact sensitivity of nitroaromatics. Equation (11.3) provides a general correlation between spark sensitivity based on RDAD instrument and impact sensitivity of nitroaromatics and nitramines. Equations (11.4) and (11.5) introduce suitable relationships between the electric spark sensitivity based on RDAD instrument and activation energy of low-temperature non-autocatalyzed thermolysis of nitramines and nitroaromatics, respectively. Equation (11.6) relates the friction sensitivity of nitramines with the activation energies of low-temperature non-autocatalyzed thermolysis. Equation (11.7) gives a linear correlation between $(x\,Gap)_{90\%\,\text{TMD}}$ and $P_{90\%\,\text{TMD}}$. Equations (11.8), (11.9), and (11.10) show the relationships between $P_{90\%\,\text{TMD}}$, $P_{95\%\,\text{TMD}}$, and $P_{98\%\,\text{TMD}}$ and the electric spark sensitivity based on RDAD instrument.

Problems

(Hint: the necessary information for some problems are given in Appendix A)

Chapter 1

Use the following equations to calculate the crystal density of the specified energetic compounds:

(1) Equation (1.1) and Table 1.1:

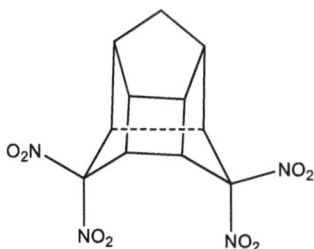

(2) Equations (1.1) and (1.5) as well as Tables 1.1 and 1.2:

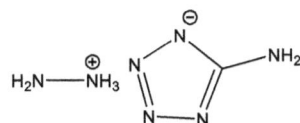

(3) Equations (1.9) and (1.9a) to (1.9g):

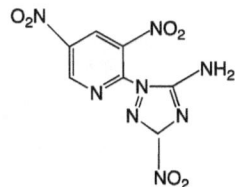

(4) Equations (1.10) and (1.10a) to (1.10e) or (1.11) or (1.12):

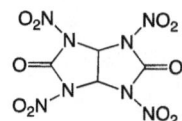

(5) Equation (1.13): $C(NO_2)_3CH_2CO_2(CH_2)_3CO_2CH_2C(NO_2)_3$

https://doi.org/10.1515/9783110740158-012

(6) Equation (1.14):

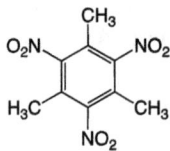

(7) Equation (1.15):

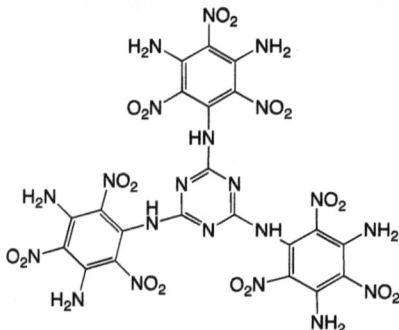

(8) Equation (1.16):

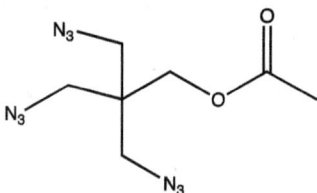

(9) Equations (1.17) and (1.18): Aminoguanidinium nitroformate
(10) Equation (1.19): 5-Azido-4-methyltetrazolium nitrate
(11) Equation (1.20):

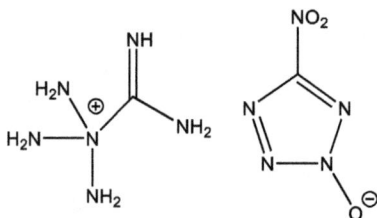

Chapter 2

Use the following equations to calculate the condensed phase heat of formation for the specified energetic compounds:

(1) Equation (2.6):

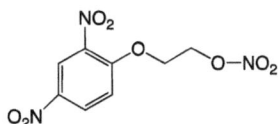

(2) Equation (2.7):

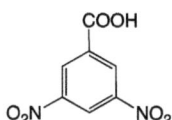

(3) Equation (2.10): $C(NO_2)_3CH_2CH_2C(=O)OCH_2C(NO_2)_3$
(4) Equation (2.11):

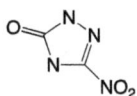

(5) Equation (2.12):

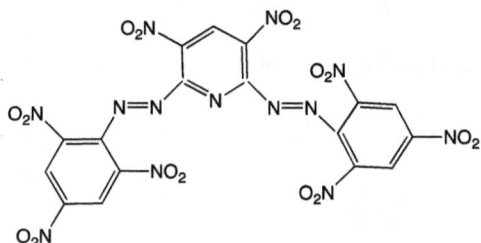

(6) Equation (2.13): 4-Nitrocinnamic acid
(7) Equation (2.14):

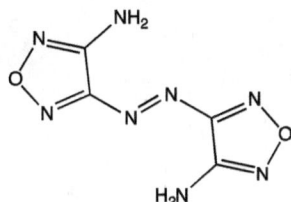

(8) Equations (2.15) and (2.16) with

$$(\Delta_f H^\theta(g))_{B3LYP} = 1077.4 \text{ kJ/mol} \quad \text{and} \quad (\Delta_f H^\theta(g))_{PM6} = 1039.5 \text{ kJ/mol},$$

respectively:

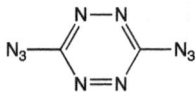

(9) Equation (2.18):

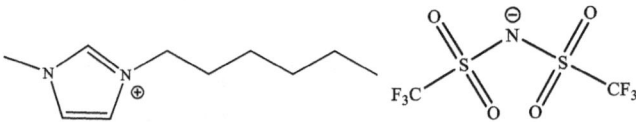

(10) Equation (2.19):

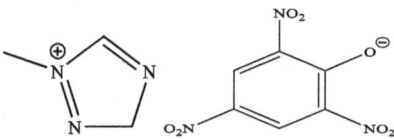

Chapter 3

Use the following equations to calculate the melting point of the specified energetic compounds:

(1) Equation (3.3):

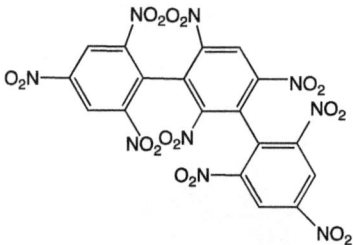

(2) Equation (3.4):

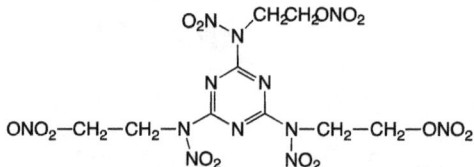

(3) Equation (3.5): $[CH_3(CH_2)_6]_4NNO_3$

(4) Equation (3.6):

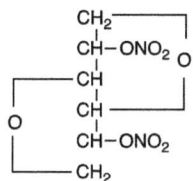

(5) Equations (3.7) to (3.9):

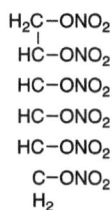

(6) Equations (3.10) to (3.12):

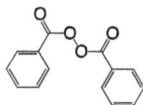

(7) Equation (3.13): 4-Chlorobenzoyl azide

(8) Equations (3.14) to (3.16):

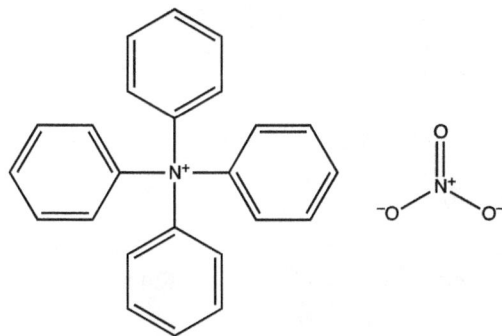

(9) Equation (3.17): (2-Tridecylpropane-1,3-diyl)dicyclohexane

(10) Equation (3.18): 1-Butyl-3-methylimidazolium trifluoromethanesulfonate

(11) Equation (3.19): 1-Nonyl-3-methylimidazolium hexafluorophosphate

Chapter 4

Use the following equations to calculate the enthalpy and entropy of fusion for the specified energetic compounds:

(1) Equation (4.1): N-Methyl-N,2,4,6-tetranitroaniline (Tetryl)
(2) Equation (4.2):

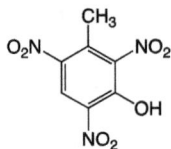

(3) Equation (4.3):

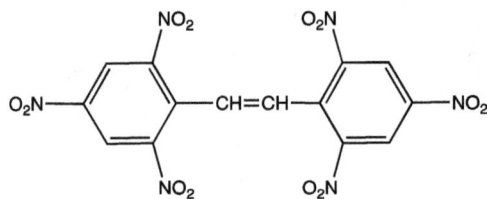

(4) Equation (4.4):

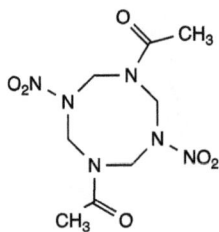

(5) Equation (4.5):

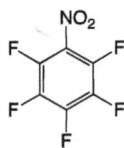

(6) Equations (4.6) and (4.7): 1,7-Diazido-2,4,6-trinitro-2,4,6-triazaheptane
(7) Equations (4.14) and (4.15):

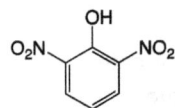

Chapter 5

Use the following equations to calculate the heat of sublimation for the specified energetic compounds:

(1) Equation (5.2): 2,2,2-Trinitro-1-phenylethane
(2) Equation (5.3):

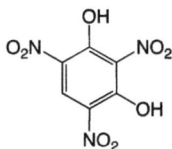

(3) Equation (5.4):

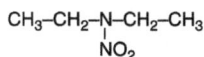

(4) Equation (5.5):

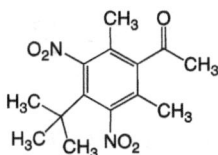

(5) Equation (5.6): N-(4-Nitrophenyl)-N-phenylamine

Chapter 6

Use the following equations to calculate the impact sensitivity for the specified energetic compounds:
(1) Equations (6.5) or (6.6): N-(2-propyl)-trinitroacetamide
(2) Equations (6.7): Bis-(2,2-dinitropropyl)-carbonate
(3) Equation (6.8): 1-Picrylimidazole
(4) Equation (6.9): 5-Nitro-1-picryl-4-picrylaminopyrazole
(5) Equation (6.10): N-Nitro-N-(3,3,3-trinitropropyl)-2,2,2-trinitroethyl carbamate
(6) Equations (6.11) and (6.12): Trinitroethyl-bis-(trinitroethoxy)-acetate
(7) Equation (6.13): Ammonium 1-hydroxy-1H,2'H-[5,5'-bitetrazol]-2'-olate

Chapter 7

Use the following equations to calculate the electric spark sensitivity of the specified energetic compounds:
(1) Equation (7.2):

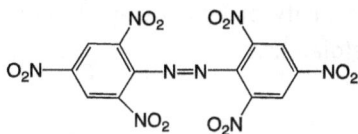

(2) Equation (7.3):

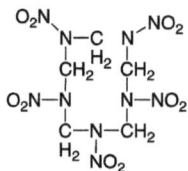

(3) Equation (7.4):

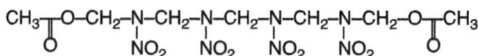

(4) Equation (7.5): 1-(2,4,6-Trinitrophenyl)-5,7-dinitrobenzotriazole (BTX)
(5) Equation (7.6) with $E_{ES,PNA}$(RDAD) = 10.61 J: 2,4,6-Tris(2,4,6-trinitrophenyl)-1,3,5-triazine (TPT)
(6) Equation (7.7): 1,3,3-Trinitroazatidine (TNAZ)
(7) Equation (7.8) with $E_{ES,PNA}$(RDAD) = 4.70 J: 2,4,6,8,10,12-Hexanitro-2,4,6,8,10,12-hexaazaisowurtzitane (HNIW)
(8) Equation (7.9): Hydroxylammonium 5,7-dinitrobenzo[d][1,2,3]triazol-1-ide

Chapter 8

Use the following equations to calculate the electric spark sensitivity of the specified energetic compounds:
(1) Equations (8.1) to (8.3): Octol-75/25
(2) Equation (8.4), if the values of ρ_0 and ρ_{TM} are 1.682 and 1.72 g/cm^3, respectively: PBX-9205
(3) Equation (8.5): Picric acid
(4) Equation (8.5): Pentolite 50/50 (PETN/TNT 50/50)

Chapter 9

(1) Use equation (9.1) to calculate the friction sensitivity of 1-(2-nitratoethylnitramino)-2,4,6-trinitrobenzene.
(2) Use equation (9.2) to calculate the friction sensitivity of hydroxylammonium 5,5'-dinitro-2H,2'H-[3,3'-bi(1,2,4- triazole)]-2,2'-bis(olate).

Chapter 10

Use the following equations to calculate the activation energies of low-temperature non-autocatalyzed thermolysis, heat of decomposition, onset temperature and deflagration temperature for the specified energetic compounds:

(1) Equation (10.1): 1,1,1-Trinitrobutane
(2) Equation (10.2):

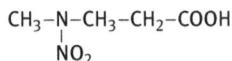

$$CH_3-N-CH_3-CH_2-COOH$$
$$\quad\;\; |$$
$$\quad\;\; NO_2$$

(3) Equation (10.3):

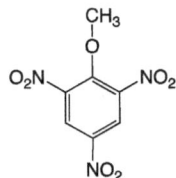

(4) Equation (10.4):

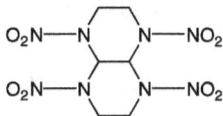

(5) Equation (10.5): 3,5-Dinitrobenzoic acid
(6) Equation (10.6): Tert-amyl peroxy-2-ethylhexyl carbonate
(7) Equation (10.7):

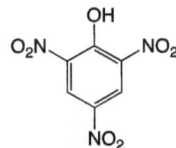

(8) Equation (10.8): Ethyl-3,3-di-(tert-amyl peroxy) butyrate
(9) Equation (10.9): 4-Azido-1-methyl-2-oxo-1,2-dihydroquinoline-3-carbaldehyde
(10) Equation (10.10):

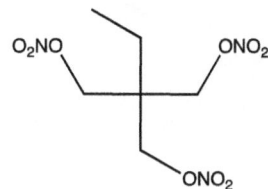

(11) Equation (10.11): 2,3-Dimethyl-1H-imidazol-3-ium bis((trifluoromethyl)sulfonyl)-amide

(12) Equation (10.12): 1-Methyl-3-(pent-2-ynyl)-1H-imidazol-3-ium azide

(13) Equation (10.13): 1-(2-(5-Nitro-2H-tetrazol-2-yl)ethyl)-1H-tetrazol-5-amine

Chapter 11

Use the following equations to calculate the relationships between the sensitivities of the specified energetic compounds:

(1) Equation (11.1) if E_a is 186.2 kJ/mol: 1,4-Dinitro-1.4-diazabutane (EDNA)

(2) Equation (11.2) if E_{IS} = 8.58 J:

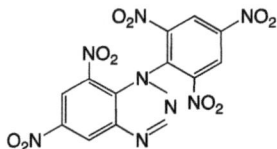

(3) Equation (11.3) if EIS = 10.20 J: 1,5-Endomethylene-3,7-dinitro-1,3,5-tetraazacyclo-octane

(4) Equation (11.4) if E_a is 178.8 kJ/mol: N,N-dimethyl-N,N-dinitroethanediamide

(5) Equation (11.5): 1,3-Dichloro-2,4,6-trinitrobenzene (DCTB)

(6) Equation (11.6) if E_a is 140 kJ/mol: 1,3,5-Trinitro-2-oxo-1,3,5-triazacyclohexane (keto-RDX)

(7) Equation (11.7) if $P_{90\% \text{ TMD}}$ is 70.38 kbar: 1,3,5-Triamino-2,4,6-trinitrobenzene (TATB)

(8) Equations (11.8), (11.9), and (11.10) if EES is 120.17 J: 1,3,5-Triamino-2,4,6-trinitrobenzene (TATB)

Answers to Problems

Chapter 1

(1) 1.704 g/cm^3
(2) 1.555 g/cm^3
(3) 1.736 g/cm^3
(4) 2.028 g/cm^3
(5) 1.669 g/cm^3
(6) 1.467 g/cm^3
(7) 1.996 g/cm^3
(8) 1.29 g/cm^3
(9) 1.755 and 1.754 g/cm^3
(10) 1.631 g/cm^3
(11) 1.722 g/cm^3

Chapter 2

(1) −77.6 kJ/mol
(2) −440.2 kJ/mol
(3) −463.2 kJ/mol
(4) −135.1 kJ/mol
(5) 664.7 kJ/mol
(6) −398.5 kJ/mol
(7) 527.0 kJ/mol
(8) 1094.4 and 1095.2 kJ/mol
(9) −2019.8 kJ/mol

Chapter 3

(1) 688 K
(2) 358.3 K
(3) 355.2 K
(4) 441.7 K
(5) 387.1 K
(6) 367 K
(7) 337 K
(8) 423.9 K
(9) 276.8 K
(10) 276.1 K
(11) 284.3 K

https://doi.org/10.1515/9783110740158-013

Chapter 4

(1) 24.88 kJ/mol
(2) 26.87 kJ/mol
(3) 39.06 kJ/mol
(4) 28.19 kJ/mol
(5) 17.00 kJ/mol
(6) 47.86 kJ/mol
(7) 1.0
(8) 757.2 K
(9) 57.9 J K^{-1} mol^{-1}

Chapter 5

(1) 88.60 kJ/mol
(2) 114.49 kJ/mol
(3) 61.77 kJ/mol
(4) 115.5 kJ/mol
(5) 125.6 kJ/mol
(6) 47.86 kJ/mol
(7) 57.9 J K^{-1} mol^{-1}

Chapter 6

(1) 95 cm
(2) 72 cm
(3) 284 cm
(4) 9 cm
(5) 9 cm
(6) 9 cm
(7) 33 J

Chapter 7

(1) 7.54 J
(2) 2.65 J
(3) 14.51 J
(4) 141.5 mJ
(5) 323.3 mJ
(6) 78.3 mJ
(7) 486.2 mJ
(8) 1380 mJ

Chapter 8

(1) $P_{90\% \text{ TMD}} = 11.34$ kbar, $P_{95\% \text{ TMD}} = 16.54$ kbar and $P_{98\% \text{ TMD}} = 22.13$ kbar
(2) 50.80 mm
(3) 2.50 mm
(4) 6.66 mm

Chapter 9

(1) 326.1 N
(2) 336 N

Chapter 10

(1) 185.41 kJ/mol
(2) 175.48 kJ/mol
(3) 113.12 kJ/mol
(4) 215.44 kJ/mol
(5) 671 kJ/mol
(6) −975.6 J/g
(7) 487.0 K
(8) 411.1 K
(9) 401.51 K
(10) 499 K
(11) 122.7 kJ/mol
(12) 398 K
(13) 493 K

Chapter 11

(1) 7.43 J
(2) 6.85 J
(3) 16.50 J
(4) 6.48 J
(5) 2.51 J
(6) 97.8 N
(7) 3.92 mm
(8) 68.24, 121.92, and 164.61 kbar

List of symbols

a	Number of carbon atoms
a'	Number of carbon atoms divided by molecular weight of explosive
a_{ani}	Number of carbon atoms in the anion
a_{cat}	Number of carbon atoms in the cation
a_i	Number of the ith atom group A_i in equation (5.2)
A_i	Contribution of the first-, second-, or third-order group of type i
A_s^+	Portion of the cation surface that has a positive electrostatic potential
A_s^-	Portion of the anion surface that has a negative electrostatic potential
AM1	Semi-empirical method
ANN	Artificial Neural Network
b	Number of hydrogen atoms
b_{ani}	Number of hydrogen atoms in the anion
b_{cat}	Number of hydrogen atoms in the cation
b'	Number of hydrogen atoms divided by molecular weight of explosive
B_i	Number of ith atoms in one mole
BRSP3	Total number of nonring, nonterminal, and branched sp^3 atoms
c	Number of nitrogen atoms
c_{ani}	Number of nitrogen atoms in the anion
c'	Number of nitrogen atoms divided by molecular weight of explosive
C	Capacitance
C_j	A correction for interactions in cycles
$C_{\text{CH}_2\text{NNO}_2 \geq 3, \text{C}(=\text{O})(\text{O or NH})}$	Presence of methylenenitramine greater than, or equal to three in cyclic nitramines or the presence of COO or CONH functional groups in equation (7.3)
CHEETAH	Thermochemical computer code
C_{De}	Contribution of specific polar groups attached to aromatic rings in equation (5.5)
C_{In}	Presence of some molecular parameters in equation (5.5)
C_{NG}	Negative contribution of structural parameter of crystal density in equation (1.14)
C_{NH_2}	Contribution of amino groups in nitroaromatics
$C_{-\text{NO}_2(-\text{ONO}_2)}$	Correcting function for the enthalpy of fusion in equation (4.4)
C_{PG}	Positive contribution of structural fragments of crystal density in equation (1.14)
$C_{\text{R,OR}}$	Presence of alkyl (−R) or alkoxy groups (−OR) in equation (7.2)
C_{SFG}	Contribution of specific functional groups in equation (3.5)

https://doi.org/10.1515/9783110740158-014

C_{SG}	Contribution of certain polar groups in equation (5.4)
C_{Shock}	Correcting function in equation (8.5)
C_{SPG}	Contribution of specific polar groups attached to an aromatic ring in equation (4.2)
C_{SSP}	Contribution of specific groups in equation (4.3)
d	Number of oxygen atoms
d_{ani}	Number of oxygen atoms in the anion
d_{cat}	Number of oxygen atoms in the cation
d_c	Critical diameter
d'	Number of oxygen atoms divided by molecular weight of explosive
d''	Number of oxygen or sulfur
D_k	A correction for atomic functional groups of kth type
DCF	Variable for decreasing the heat content of an energetic compound in equation (2.15)
DF	Variable for decreasing the heat content of an energetic compound in equation (2.14)
DFT	Density functional theory
DMP	Diminishing intermolecular interaction for decreasing crystal density in equation (1.15)
DSC	Differential scanning calorimetry
DT	Deflagration temperature
DTA	Differential thermal analysis
$DSSP$	Decreasing effects of melting point of some specific structural features in equation (3.4)
$DSSPH$	Decreasing sensitivity structural parameter in equation (6.10)
e	Number of fluorine atoms
e_{ani}	Number of fluorine atoms in the anion
E	Variable for decreasing heat content of an energetic compound in equation (2.7)
E_a	Activation energy of low-temperature non-autocatalyzed thermolysis
$E_{a,IL}$	Activation energy of thermolysis of the desired ionic liquid or salt
$E_{a,IL}^+$	Correcting function for increasing activation energy of thermolysis of the desired ionic liquid or salt in equation (10.11)
$E_{a,IL}^-$	Correcting function for decreasing activation energy of thermolysis of the desired ionic liquid or salt in equation (10.11)
E_{cor}^+	Correcting function for increasing E_{ES} in equation (11.2)
E_{cor}^-	Correcting function for decreasing E_{ES} in equation (11.2)
E_D	Decreasing structural parameter of crystal density in equation (1.13)
E_{ES}	Electrostatic or electric spark sensitivity

$E_{ES,NTA}$(ESZ KTTV) Electric spark sensitivity of nitramines based on the ESZ KTTV system

$E_{ES,PNA}$(ESZ KTTV) Electric spark sensitivity of polynitro arenes based on the ESZ KTTV system

$E_{ES,NTA}$(RDAD) Electric spark sensitivity based on the RDAD system

$E_{ES,PNA}$(RDAD) Electrostatic sensitivity based on the RDAD system for polynitro arenes

$E^+_{ES,NTA}$ Correcting function in equation (7.6)

$E^-_{ES,NTA}$ Correcting function that can decrease the predicted results on the basis of the RDAD system in equation (7.8) for nitramines

$E^+_{ES,PNA}$ Correcting function in equation (7.5)

$E^{+'}_{ES,PNA}$ Correcting function that can increase the predicted results on the basis of the RDAD system in equation (7.6)

$E^-_{ES,PNA}$ Correcting function that can decrease the predicted results on the basis of the RDAD system in equation (7.6) for polynitro arenes

E^{++}_{IS} Correcting function for increasing E_{IS} in equation (11.1)

E^{--}_{IS} Correcting function for decreasing E_{IS} in equation (11.1)

$E^+_{ES,Ar}$ Correcting function for increasing E_{ES} in equation (11.5)

$E^-_{ES,Ar}$ Correcting function for decreasing E_{ES} in equation (11.5)

$E^+_{ES,corr}$ Increasing electric spark sensitivity factor in equation (11.4)

$E^+_{ES,NAr,NiA}$ Correcting function for increasing E_{ES} in equation (11.3)

$E^-_{ES,NAr,NiA}$ Correcting function for decreasing E_{ES} in equation (11.3)

E_I Increasing structural parameter of crystal density in equation (1.13)

E_{IS} Impact sensitivity in J

E^+_{IS} Correcting function for increasing E_{ES} in equation (7.4)

$E_{N\ excess}$ Existence of excess nitrogen atoms for cyclic nitramines containing excess nitrogen atoms beside $-NNO_2$ and $-NO_2$

$E^+_{non-add}$ Contribution of non-additive structural parameters for increasing E_a in equation (10.4)

$E^-_{non-add}$ Contribution of non-additive structural parameters for decreasing E_a in equation (10.4)

$E^0_{\alpha CH/NNO_2}$ A parameter that indicates the existence of α-C–H in nitroaromatic compounds or the N–NO$_2$ functional group

ES_{IL} (mJ) Sensitivity toward electrical discharge of a desired energetic ionic compound in mJ

ES^+_{IL} Correcting function for increasing electric spark of quaternary ammonium-based energetic ionic liquids or salts based on ESZ KTTV

ES^-_{IL} Correcting function for decreasing electric spark of quaternary ammonium-based energetic ionic liquids or salts based on ESZ KTTV

EDPHT	Computer code based on empirical correlations
EILoS	Energetic ionic liquids or salts
E–P–S	Evans–Polanyi–Semenov
ESZ KTTV	An instrument for measuring electric spark sensitivity
EXPLO5	Thermochemical computer code
f	Number of chlorine atoms
f_{ani}	Number of chlorine atoms in the anion
F	Variable for increasing the heat content of an energetic compound in equation (2.7)
F^+	Existence of specific molecular moieties for increasing value of $(\log H_{50})_{core}$
F^-	Existence of specific molecular moieties for decreasing value of $(\log H_{50})_{core}$
$F_{attract}$	Attractive intermolecular forces in equation (5.6)
$F^+_{non\text{-}add}$	Contribution of non-additive structural parameters for increasing the deflagration temperature of equation (10.10)
$F^-_{non\text{-}add}$	Contribution of non-additive structural parameters for decreasing the deflagration temperature of equation (10.10)
F_{repul}	Repulsive intermolecular forces in equation (5.6)
FS	Friction sensitivity
FS^+	Correcting function for increasing friction sensitivity in equation (11.6)
FS^-	Correcting function for decreasing friction sensitivity in equation (11.6)
$FS_{IL}(N)$	Friction sensitivity of quaternary ammonium-based energetic ionic liquids in newton
FS^+_{IL}	Correcting function for increasing friction sensitivity of ionic liquids in equation (9.2)
FS^-_{IL}	Correcting function for decreasing friction sensitivity of ionic liquids in equation (9.2)
g	Number of aluminum atoms
g_{ani}	Number of aluminum atoms in the anion
g_j	Number of the special group G_j in equation (5.2)
G_{50}	Barrier thickness required to inhibit detonation in the test explosive half the time for gap test
GA	Genetic algorithm
GAV	Group additivity values
GIPF	Generalized interaction property function
H_{50}	Impact drop height
l_{ani}	Number of boron atoms in anion
$H_{fus,i}$	Contribution of molecular fragment or group i to the enthalpy of fusion in equation (4.1)

h_{ani}	Number of sulfur atoms in anion
ICF	Variable for increasing heat content of an energetic compound in equation (2.15)
IF	Variable for increasing heat content of an energetic compound in equation (2.14)
IMP	Increment of intermolecular interaction for increasing crystal density in equation (1.15)
IPF	Inefficient packing factor in equation (3.13)
IS_{IL}^+	Correction function of friction sensitivity that depends on stabilizing structural parameters in cations or anions
IS_{IL}^-	Correction function of friction sensitivity that depends on destabilizing structural parameters in cations or anions
ISPBKW	Computer code for calculation of the specific impulse using BKW-EOS
ISSP	Increasing effects of melting point of some specific structural features in equation (3.4)
ISSPH	Increasing sensitivity structural parameter in equation (6.10)
k	Number of halogen atoms
L_j	Number of cycles of jth order in one mole
l_{ani}	Number of boron atoms in anion
LANL	Los Alamos National Laboratory
LINSP3	Number of nonring, nonterminal, and nonbranched sp^3 atoms
M	Molecular mass of the molecule in g/molecule
$m_{cyc}^{>6}$	Cyclic nitramines with rings which are larger than six membered rings and which contain only carbon and nitrogen atoms
MD	Molecular Dynamics
MLR	Multilinear regression
MESP	Molecular surface electrostatic potential
MM (MM2, MM3, ...)	Molecular mechanics methods
Mw	Molecular weight of the desired energetic compound
Mw'	Molecular weight of the desired energetic compound under certain conditions in equations (5.5) and (5.6)
Mw_{cat}	Molecular weights of cation
Mw_{ani}	Molecular weights of anion
$n_{c,i}$	Occurrence of the groups i in the cation of the desired ionic liquid
$n_{a,j}$	Occurrence of the groups j in the anion of the desired ionic liquid
n_{EDNA}	Number of N,N'-(ethane-1,2-diyl)dinitramide (EDNA) moieties ($O_2NNCH_xCH_xNNO_2$) in cyclic nitaramines
n_{H_2O}	Number of the H_2O molecules in the crystal of the salt
N_j	Number of atoms in a cycle of jth order
N_k	Number of atoms in the group of k type
n_x	The number of x

n'_x	Number of x divided by molecular weight of explosive
n'_{Ar}	The number of aromatic rings under certain conditions for equation (2.7)
$n_{NNO_2}^{>3,linear}$	Number of $-NNO_2$ groups for those acyclic linear nitramines containing more than three nitramine groups in equation (4.4)
NASA-CEC-71	Computer program for calculation of complex chemical equilibrium compositions
NLR	Non-linear regression
NSWC	Naval Surface Warfare Center
OB_{100}	Oxygen balance
p	Number of cations
P_{FS}^+	Correcting parameter for increasing friction sensitivity in equation (9.1)
P_{FS}^-	Correcting parameter for decreasing friction sensitivity in equation (9.1)
P_x	Presence of x
$P_{90\%\,TMD}$	Pressure in kbar required to initiate material pressed to 90 % of theoretical maximum density
$P_{95\%\,TMD}$	Pressure in kbar required to initiate material pressed to 95 % of theoretical maximum density
$P_{98\%\,TMD}$	Pressure in kbar required to initiate material pressed to 98 % of theoretical maximum density
$P_{>5}$	Existence of cyclic nitramines that contain more than five member ring as well as cage nitramines
PLS	Partial least squares
PM3, PM6, PM7, ...	Semi-empirical methods
q	Number of anions
Q_{corr}	The corrected heats of detonation on the basis of Kamlet's method
QM	Quantum mechanical method
QSPR	Quantitative Structure–Property Relationships
R_k	Number of groups of k-type in one mole
RDAD	An instrument for measuring electric spark sensitivity
RING	Number of single, fused, or conjugated ring systems
ROT	Extra entropy produced by freely rotating sp^3 atoms
V_m	Volume inside the 0.001 a. u. isosurface of electron density surrounding the molecule
SA	Molecular surface area
SADT	Self-accelerating decomposition temperature
SPG	Contribution of specific polar groups of melting point in equation (3.13)
SMM	Soviet manometric method
SP2	Number of nonring, and nonterminal sp^2 atoms

SPARC	SPARC Performs Automated Reasoning in Chemistry
SSP_i	Contribution of structural parameters in equation (3.13)
SVM	Support vector machine
T^+	Increasing parameter of melting point in equation (3.6)
T^-	Decreasing parameter of melting point in equation (3.6)
T_{add}	Additive function of melting point of equation (3.7)
$T_{add,elem}$	Contribution of the elemental composition as an additive part in equation (3.14)
T_{core}	Core contribution of elemental composition for prediction of melting point in equation (3.10)
$T_{corr,strut}$	Non-additive part of the melting point in equation (3.14)
$T_{correcting}$	Correcting function of melting point in equation (3.10)
T_{dmax}	Temperature of maximum mass loss
T_{dmax}^+	Correcting function for increasing T_{dmax} in equation (10.9)
T_{dmax}^-	Correcting function for decreasing T_{dmax} in equation (10.9)
T_m	Melting point
$T_{c,i}$	Contribution of cation groups i for prediction of melting point
$T_{a,j}$	Contribution of anion groups j for prediction of melting point
$T_{m,peroxide}$	Melting point of peroxide compound
$T_{m,cyc\ hyd}^+$	Correction term for increment of melting point of cyclic saturated and unsaturated hydrocarbons in equation (3.17)
$T_{m,cyc\ hyd}^-$	Correction term for decreasing melting point of cyclic saturated and unsaturated hydrocarbons in equation (3.17)
$T_{m,IL}^+$	Positive adjusting function in equation (3.19)
$T_{m,IL}^-$	Negative adjusting function in equation (3.19)
$T_{non-add}$	Non-additive function of melting point of equation (3.7)
$T_{o,p}$	A parameter of equation (3.3) that can be applied in disubstituted benzene ring
T_{onset}	Onset temperature
$T_{onset,azole}$	Onset decomposition temperature of the desired azole based energetic compound
$T_{onset,azole}^+$	Correcting parameter for increasing onset decomposition temperature in equation (10.13)
$T_{onset,azole}^-$	Correcting parameter for decreasing onset decomposition temperature in equation (10.13)
$T_{onset,IL}$	Onset decomposition temperature of imidazolium based energetic ionic liquids or salts
$T_{onset,IL}^+$	Correcting parameter for increasing onset decomposition temperature in equation (10.12)
$T_{onset,IL}^-$	Correcting parameter for decreasing onset decomposition temperature in equation (10.12)
T_{SFG}	Contribution of a specific functional group in equation (3.3)

T^+_{struc}	Increasing structural parameters of melting point in equation (3.16)
T^-_{struc}	Decreasing structural parameters of melting point in equation (3.16)
T_{PC}	Increasing non-additive parameter of melting point in equation (3.9)
T_{NC}	Decreasing non-additive parameter of melting point in equation (3.9)
TET	Correcting parmeter in equation (1.20) for tetrazole salts containing 1-N-oxide and 2-N-oxide fragment
TGA	Thermogravimetry
TMD	Theoretical maximum density
U	Voltage
UPPER	Unified Physical Property Estimation Relationships
V	Total volume
V_s	Electrostatic potential on the surface
V^+	Effective volume of cation
V^-	Effective volume of anion
$\bar{V}^+_s$	Average of the positive values of V_s
$\bar{V}^-_s$	Average of the negative values of V_s
$Void_{\text{theo}}$	Theoretical calculated percent void
$(x\,Gap)_{90\,\%\,\text{TMD}}$	Thickness of aluminum in mm at 90 % of TMD
$(\log H_{50})_{\text{core}}$	Impact drop height on the basis of elemental composition in equation (6.12)
ΔH_{decom}	Heat of decomposition in kJ/mol
$\Delta H'_{\text{decom}}$	Heat of decomposition in J/g
$\Delta H^+_{\text{decom}}$	Contribution of non-additive structural parameters for increasing ΔH_{decom} in equation (10.5)
$\Delta H^-_{\text{decom}}$	Contribution of non-additive structural parameters for decreasing ΔH_{decom} in equation (10.5)
$\Delta H_{\text{Inc,fus}}$	Contribution of structural parameter for increasing enthalpy of fusion in equation (4.5)
$\Delta H_{\text{Dec,fus}}$	Contribution of structural parameter for decreasing enthalpy of fusion in equation (4.5)
$(\Delta_{\text{fus}}H)_{\text{add}}$	Additive contribution of elemental composition in equation (4.6)
$(\Delta_{\text{fus}}H)^{\text{Inc}}_{\text{Non-add}}$	Non-additive contribution for increasing effects of specific groups in equation (4.6)
$(\Delta_{\text{fus}}H)^{\text{Dec}}_{\text{Non-add}}$	Non-additive contribution for decreasing effects of specific groups in equation (4.6)
$\Delta_f H^{\theta}(c)[EILoS]$	Condensed phase heat of formation of energetic ionic liquids or salts

$\Delta_f H^\theta(c)[IMILoS]$	Condensed phase heat of formation of imidazolium-based ionic liquids or salts
$\Delta_f H^\theta(c)[TAILoS]$	Condensed phase heat of formation of of triazolium-based energetic ionic liquids or salts
$\Delta_f H(g)$	Gas phase heat of formation
$\Delta_f H^\theta(g)$	Standard gas phase heat of formation
$\Delta_f H^\theta(l)$	Standard liquid phase heat of formation
$\Delta_f H(s)$	Solid phase heat of formation
$\Delta_f H^\theta(s)$	Standard solid phase heat of formation
$\Delta_f H^\theta_{\text{IEC}}$	Correcting function for increasing heat content of an energetic compound in equation (2.12)
$\Delta_f H^\theta_{\text{DEC}}$	Correcting function for decreasing heat content of an energetic compound in equation (2.12)
$\Delta_f H^\theta_{\text{add,DHC}}$	Correcting additive function for decreasing heat content of an energetic compound in equation (2.13)
$\Delta_f H^\theta_{\text{non-add,DHC}}$	Correcting non-additive function for decreasing heat content of an energetic compound in equation (2.13)
$\Delta_f H^\theta_{\text{add,IHC}}$	Correcting additive function for increasing heat content of an energetic compound in equation (2.13)
$\Delta_f H^\theta_{\text{non-add,IHC}}$	Correcting non-additive function for increasing heat content of an energetic compound in equation (2.13)
$\Delta_f H^\theta_{\text{Inc}}[IMILoS]$	Increasing and decreasing heat contents in imidazolium-based ionic liquids or salts in equation (2.18)
$\Delta_f H^\theta_{\text{Dec}}[IMILoS]$	Decreasing heat contents in imidazolium-based ionic liquids or salts in equation (2.18)
$\Delta_f H^\theta_{\text{Inc}}[TAILoS]$	Increasing function in the triazolium-based energetic ionic liquids or salts in equation (2.19)
$\Delta_f H^\theta_{\text{Dec}}[TAILoS]$	Decreasing function in the triazolium-based energetic ionic liquids or salts in equation (2.19)
$\Delta_f H^\theta(g)$	Standard heat of formation of a specific compound in the gas phase
$[\Delta_f H^\theta(g)]_{\text{B3LYP/6-31G*}}$	Gas phase heat of formation in kJ/mol using the B3LYP/6-31G* method
$[\Delta_f H^\theta(g)]_{\text{PM3}}$	Gas phase heat of formation in kJ/mol using the PM3 method
$[\Delta_f H^\theta(g)]_{\text{PM6}}$	Gas phase heat of formation in kJ/mol using the PM6 method
$\Delta_f H^\theta(c)$	Standard heat of formation of a specific compound in the condensed phase (solid or liquid)
$\Delta_f H^\theta(c)[EILoS]$	Condensed phase heat of formation of EILoS
$\Delta_f H^\theta(c)[IMILoS]$	Condensed phase heat of formation of imidazolium-based ionic liquids or salts
$\Delta_f H^\theta(c)[TAILoS]$	Condensed phase heat of formation of triazolium-based energetic ionic liquids or salts

$\sum_i \Delta_f H_i^{\theta}(g)$[cation]	Sum of gas-phase heats of formation of cations $\sum_j \Delta_f H_j^{\theta}(g)$[anion] Sum of gas-phase heats of formation of anions
$\Delta_{fus}H$	Enthalpy or heat of fusion
$\Delta_{sub}H$	Heat of sublimation
$\Delta_{sub}H^{\theta}$	Standard heat of sublimation
$\Delta_{vap}H^{\theta}$	Standard heat of vaporization.
$\Delta_{fus}S$	Entropy of fusion
ΔH_{PT}	Heat of phase transition
$\Delta_{fus}S_{add}$	Additive contribution of entropy of fusion in equation (4.14)
$\Delta_{fus}S_{non\text{-}add}$	Non-additive contribution of entropy of fusion in equation (4.14)
ρ	Crystal density in g/cm^3
ρ'	Uncorrected crystal density in g/cm^3
ρ^+	Increasing crystal density parameter in equation (1.18)
ρ^-	Decreasing crystal density parameter in equation (1.18)
ρ_0	Loading density
ρ_{azide}^+	Increasing crystal density parameter of azide compounds in equation (1.16)
ρ_{azide}^-	Decreasing crystal density parameter of azide compounds in equation (1.16)
$\rho_{Tetrazole\text{-}N\text{-}oxide}^+$	Positive non-additive structural parameter in equation (1.20)
$\rho_{Tetrazole\text{-}N\text{-}oxide}^-$	Negative non-additive structural parameter in equation (1.20)
$\rho_{tetrazolium\ nitrate}$	Decreasing structural parameter in tetrazolium nitrate salts
ρ_{TM}	Theoretical maximum density
σ	Molecular rotational symmetry
σ'	Pseudosymmetry
σ_{tot}^2	Total variance of the electrostatic potential on the 0.001 a. u. molecular surface
ν	Degree of balance between the positive and negative potentials on the molecular surface
ε	Molecular eccentricity
ε_{ar}	Aromatic eccentricity
ε_{al}	Aliphatic eccentricity
ϕ	Flexibility number
Π	Average deviation of electrostatic potential
$(C\text{-}N(NO_2)\text{-}C)_{pure}$	Contribution of C–N(NO$_2$)–C linkage in pure nitramines
P_{FS}^+	Correcting parameter of increasing friction sensitivity in equation (9.1)
P_{FS}^-	Correcting parameter of decreasing friction sensitivity in equation (9.1)
$|n_{TNB} - 2|$	Absolute value of the number of 1,3,5-trinitrobenzene minus two in equation (10.7)

λ' Positive and negative contributions of various structural parameters to obtain more reliable T_{onset} in equation (10.8)

λ_{sym} Peroxides containing the same fragments attached to the $-O-O-$ bond in equation (10.8)

A Glossary of compound names and heats of formation for pure as well as composite explosives

Abbreviation	Full name or composition	Chemical formula	$\Delta_f H^\theta$ (c) (kJ mol^{-1})
ABH	Azobis (2,2′,4,4′,6,6′-hexanitrobiphenyl)	$C_{24}H_6N_{14}O_{24}$	485.34 [459]
Alex 20	44/32/20/4 RDX/TNT/Al/Wax	$C_{1.783}H_{2.469}N_{1.613}O_{2.039}Al_{0.7335}$	−7.61
Alex 32	37/28/31/4 RDX/TNT/Al/Wax	$C_{1.647}H_{2.093}N_{1.365}O_{1.744}Al_{1.142}$	−9.33
AMATEX-20	42/20/38 AN/RDX/TNT	$C_{1.44}H_{1.38}N_{1.04}O_{1.54}(AN)_{0.53}$	−95.77
AMATEX-40	21/41/38 AN/RDX/TNT	$C_{1.73}H_{1.95}N_{1.61}O_{2.11}(AN)_{0.26}$	−197.49
AMATOL80/20	80/20 AN/TNT	$C_{0.62}H_{0.44}N_{0.26}O_{0.53}(AN)_1$	−371.25
AN	Ammonium nitrate	NH_4NO_3 or $H_4N_2O_3$	−365.14 [375]
AN/Al (90/10)	–	$Al_{0.37}(AN)_{1.125}$ or $H_{4.5}N_{2.25}O_{3.37}Al_{0.37}$	−412.42
AN/Al (80/20)	–	$Al_{0.74}(AN)_1$ or $H_4N_2O_3Al_{0.74}$	−368.32
AN/Al (70/30)	–	$Al_{1.11}(AN)_{0.875}$ or $H_{3.5}N_{1.75}O_{2.62}Al_{1.11}$	−324.55
BTF	Benzotris(1,2,5-oxadiazole-1-oxide)	$C_6N_6O_6$	602.50 [459]
COMP A-3	91/9 RDX/WAX	$C_{1.87}H_{3.74}N_{2.46}O_{2.46}$	11.88 [459]
COMP B	63/36/1 RDX/TNT/wax	$C_{2.03}H_{2.64}N_{2.18}O_{2.67}$	5.36 [375]
COMP C-3	77/4/10/5/1/3 RDX/TNT/DNT/MNT/NC/Tetryl	$C_{1.90}H_{2.83}N_{2.34}O_{2.60}$	−26.99 [459]
COMP C-4	91/5.3/2.1/1.6 RDX/TNT/MNT/NC	$C_{1.82}H_{3.54}N_{2.46}O_{2.51}$	13.93 [459]
Cyclotol-50/50	50/50 RDX/TNT	$C_{2.22}H_{2.45}N_{2.01}O_{2.67}$	0.04
Cyclotol-60/40 (or COMP B-3)	60/40 RDX/TNT	$C_{2.04}H_{2.50}N_{2.15}O_{2.68}$	4.81 [375]
Cyclotol-65/35	65/35 RDX/TNT	$C_{1.96}H_{2.53}N_{2.22}O_{2.68}$	8.33
Cyclotol-70/30	70/30 RDX/TNT	$C_{1.87}H_{2.56}N_{2.29}O_{2.68}$	11.13
Cyclotol-75/25	75/25 RDX/TNT	$C_{1.78}H_{2.58}N_{2.36}O_{2.69}$	13.4 [375]
Cyclotol-77/23	77/23 RDX/TNT	$C_{1.75}H_{2.59}N_{2.38}O_{2.69}$	14.98
Cyclotol-78/22	78/22 RDX/TNT	$C_{1.73}H_{2.59}N_{2.40}O_{2.69}$	15.52
DATB	1,3-Diamino-2,4,6-trinitrobenzene	$C_6H_5N_5O_6$	−98.74 [459]
Destex	74.766/18.691/4.672/1.869 TNT/Al/Wax/Graphite	$C_{2.791}H_{2.3121}N_{0.987}O_{1.975}Al_{0.6930}$	−34.39
DIPAM (Dipicramide)	2,2′,4,4′,6,6′-Hexanitro-[1,1-biphenyl]-3,3′-diamine	$C_{12}H_6N_8O_{12}$	−14.90 [460]
DIPAM (Dipicramide)	2,2′,4,4′,6,6′-Hexanitro-[1,1-biphenyl]-3,3′-diamine	$C_{12}H_6N_8O_{12}$	−28.45 [459]
EXP D	Ammonium picrate or Explosive D	$C_6H_6N_4O_7$	−393.30 [459]
EDC-11	64/4/30/1/1 HMX/RDX/TNT/Wax/Trylene	$C_{1.986}H_{2.78}N_{2.23}O_{2.63}$	4.52
EDC-24	95/5 HMX/Wax	$C_{1.64}H_{3.29}N_{2.57}O_{2.57}$	18.28

https://doi.org/10.1515/9783110740158-015

Abbreviation	Full name or composition	Chemical formula	$\Delta_f H^\theta$ (c) (kJ mol^{-1})
HBX-3	31/29/35/5/0.5 RDX/TNT/AL/WAX/CaCl$_2$	$C_{1.66}H_{2.18}N_{1.21}O_{1.21}O_{1.60}Al_{1.29}Ca_{0.005}Cl_{0.009}$	−8.71 [461]
HMX	Cyclotetramethylenetetranitramine	$C_4H_8N_8O_8$	74.98 [459]
HMX/Al (80/20)	—	$C_{1.08}H_{2.16}N_{2.16}O_{2.16}Al_{0.715}$	20.21
HMX/Al (70/30)	—	$C_{0.944}H_{1.888}N_{1.888}O_{1.888}Al_{1.11}$	17.66
HMX/Al (60/40)	—	$C_{0.812}H_{1.624}N_{1.624}O_{1.624}Al_{1.483}$	15.19
HMX/Exon (90.54/9.46)	—	$C_{1.43}H_{2.61}N_{2.47}O_{2.47}F_{0.15}Cl_{0.10}$	−1026.80
HNAB	2,2′,4,4′,6,6′-Hexanitroazobenzene	$C_{12}H_4N_8O_{12}$	284.09 [459]
Liquid TNT	—	$C_7H_5N_3O_6$	−53.26
LX-04	85/15 HMX/Viton	$C_{5.485}H_{9.2229}N_8O_8F_{1.747}$	−89.96 [375]
LX-07	90/10 HMX/Viton	$C_{1.48}H_{2.62}N_{2.43}O_{2.43}F_{0.35}$	−51.46 [375]
LX-09	93/4.6/2.4 HMX/DNPA/FEFO	$C_{1.43}H_{2.74}N_{2.59}O_{2.72}F_{0.02}$	8.38 [375]
LX-10	95/5 HMX/Viton	$C_{1.42}H_{2.66}N_{2.57}O_{2.57}F_{0.17}$	−13.14 [375]
LX-11	80/20 HMX/Viton	$C_{1.61}H_{2.53}N_{2.16}O_{2.16}F_{0.70}$	−128.57 [375]
LX-14	95.5/4.5 HMX/Estane 5702-F1	$C_{1.52}H_{2.92}N_{2.59}O_{2.66}$	6.28 [459]
LX-15	95/5 HNS-I/Kel-F 800	$C_{3.05}H_{1.29}N_{1.27}O_{2.53}Cl_{0.04}F_{0.3}$	−18.16 [375]
LX-17	92.5/7.5 TATB/Kel-F 800	$C_{2.29}H_{2.18}N_{2.15}O_{2.15}Cl_{0.054}F_{0.2}$	−100.58 [375]
MEN-II	72.2/23.4/4.4 Nitromethane/Methanol/Ethylene diamine	$C_{2.06}H_{7.06}N_{1.33}O_{3.10}$	−310.87 [375]
MINOL-2	40/40/20 AN/TNT/Al	$C_{1.23}H_{0.88}N_{0.53}O_{1.06}Al_{0.74}(AN)_{0.5}$	−194.26 [251]
NM	Nitromethane	$C_1H_3N_1O_2$	−112.97 [459]
NONA	2,2′,2″,4,4′,4″,6,6′,6″-Nonanitro-m-terphenyl	$C_{18}H_5N_9O_{18}$	114.64 [459]
NQ	Nitroguanidine	$CH_4N_4O_2$	−92.47 [251]
NM/UP (60/40)	60/40 Nitromethane/UP; UP = 90/10 CO(NH$_2$)$_2$ HClO$_4$/H$_2$O	$C_{1.207}H_{4.5135}N_{1.432}O_{3.309}Cl_{0.2341}$	11.51
Octol-76/23	76.3/23.7 HMX/TNT	$C_{1.76}H_{2.58}N_{2.37}O_{2.69}$	12.76
Octol-75/25	75/25 HMX/TNT	$C_{1.78}H_{2.58}N_{2.36}O_{2.69}$	11.63 [375]
Octol-60/40	60/40 HMX/TNT	$C_{2.04}H_{2.50}N_{2.15}O_{2.68}$	4.14
PBX-9007	90/9.1/0.5/0.4 RDX/Polystyrene/DOP/Resin	$C_{1.97}H_{3.22}N_{2.43}O_{2.44}$	29.83 [459]
PBX-9010	90/10 RDX/Kel-F	$C_{1.39}H_{2.43}N_{2.43}O_{2.43}Cl_{0.09}F_{0.26}$	−32.93 [375]
PBX-9011	90/10 HMX/Estane	$C_{1.73}H_{3.18}N_{2.45}O_{2.61}$	−16.95 [459]
PBX-9205	92/6/2 RDX/Polystyrene/DOP	$C_{1.83}H_{3.14}N_{2.49}O_{2.51}$	24.31 [459]
PBX-9407	94/6 RDX/Exon 461	$C_{1.41}H_{2.66}N_{2.54}O_{2.54}Cl_{0.07}F_{0.09}$	3.39 [375]

Abbreviation	Full name or composition	Chemical formula	$\Delta_f H^\theta$(c) (kJ mol^{-1})
PBX-9501	95/2.5/2.5 HMX/Estane/BDNPA-F	$C_{1.47}H_{2.86}N_{2.60}O_{2.69}$	9.62 [459]
PBX-9502	95/5 TATB/Kel-F 800	$C_{2.3}H_{2.23}N_{2.21}O_{2.21}Cl_{0.04}F_{0.13}$	-87.15 [375]
PBX-9503	15/80/5 HMX/TATB/KEL-F 800	$C_{2.16}H_{2.28}N_{2.26}O_{2.26}Cl_{0.038}$	-74.01 [375]
PBXC-9	75/20/5 HMX/Al/Viton	$C_{1.15}H_{2.14}N_{2.03}O_{2.03}F_{0.17}Al_{0.74}$	113.01
PBXC-116	86/14 RDX/Binder	$C_{1.968}H_{3.7463}N_{2.356}O_{2.4744}$	4.52
PBXC-117	71/17/12 RDX/Al/Binder	$C_{1.65}H_{3.1378}N_{1.946}O_{2.048}Al_{0.6303}$	-65.56
PBXC-119	82/18 HMX/Binder	$C_{1.817}H_{4.1073}N_{2.2149}O_{2.6880}$	18.28
Pentolite-50/50	50/50 TNT/PETN	$C_{2.33}H_{2.37}N_{1.29}O_{3.22}$	-100.01
PETN	Pentaerythritol tetranitrate	$C_5H_8N_4O_{12}$	-538.48 [375]
PF	1-Fluoro-2,4,6-trinitrobenzene	$C_6H_2N_3O_6F$	-224.72
RDX	Cyclomethylenetrinitramine	$C_3H_6N_6O_6$	61.55 [375]
RDX/Al (90/10)	—	$C_{1.215}H_{2.43}N_{2.43}O_{2.43}Al_{0.371}$	24.89
RDX/Al (80/20)	—	$C_{1.081}H_{2.161}N_{2.161}O_{2.161}Al_{0.715}$	22.13
RDX/Al (70/30)	—	$C_{0.945}H_{1.89}N_{1.89}O_{1.89}Al_{1.11}$	19.37
RDX/Al (60/40)	—	$C_{0.81}H_{1.62}N_{1.62}O_{1.62}Al_{1.483}$	16.61
RDX/Al (50/50)	—	$C_{0.675}H_{1.35}N_{1.35}O_{1.35}Al_{1.853}$	13.85
RDX/TFNA (65/35)	—	$C_{1.54}H_{2.64}N_{2.2}O_{2.49}F_{0.44}$	-823.83
RDX/Exon (90.1/9.9)	—	$C_{1.44}H_{2.6}N_{2.44}O_{2.44}F_{0.17}Cl_{0.11}$	-195.48
TATB	1,3,5-Triamino-2,4,6-trinitrobenzene	$C_6H_6N_6O_6$	-154.18 [375]
TATB/HMX/Kel-F (45/45/10)	—	$C_{1.88}H_{2.37}N_{2.26}O_{2.26}F_{0.28}Cl_{0.06}$	-478
Tetryl	N-Methyl-N-nitro-2,4,6-trinitroaniline	$C_7H_5N_5O_8$	19.54 [375]
TFENA	2,2,2-Trifluoroethylnitramine	$C_2H_3N_2O_2F_3$	-694.54
TFET	2,4,6-Trinitrophenyl-2,2,2-trifluoroethylnitramine	$C_8H_4N_5O_8F_3$	-576.8
TNT	2,4,6-Trinitrotoluene	$C_7H_5N_3O_6$	-67.07 [54]
TNTAB	Trinitrotriazidobenzene	$C_6N_{12}O_6$	1129.68 [459]
TNT/Al (89.4/10.6)	—	$C_{2.756}H_{1.969}N_{1.181}O_{2.362}Al_{0.393}$	-24.73
TNT/Al (78.3/21.7)	—	$C_{2.414}H_{1.724}N_{1.034}O_{2.069}Al_{0.804}$	-21.63
Toluene/Nitromethane(14.5/85.5)	—	$C_{2.503}H_{5.461}N_{1.4006}O_{2.8013}$	-160.71
Torpex	42/40/18 RDX/TNT/Al	$C_{1.8}H_{2.015}N_{1.663}O_{2.191}Al_{0.6674}$	-0.17
Tritonal	80/20 TNT/Al	$C_{2.465}H_{1.76}N_{1.06}O_{2.11}Al_{0.741}$	-23.64

B Calculation of the gas phase standard enthalpies of formation

The gas phase standard enthalpies of formation at 298 K, for $C_aH_bN_cO_d$ species, can be obtained by correcting the standard enthalpies of formation at 0 K (equation (B.1)) [273, 462].

$$\Delta_f H^\theta(g)(C_aH_bN_cO_d, 298\ K) = \Delta_f H^\theta(g)(C_aH_bN_cO_d, 0\ K)$$
$$+ [H^\theta(C_aH_bN_cO_d, 298\ K) - H^\theta(C_aH_bN_cO_d, 0\ K)]$$
$$- a \cdot [H^\theta(C, 298\ K) - H^\theta(C, 0\ K)]_{St}$$
$$- b \cdot [H^\theta(H, 298\ K) - H^\theta(H, 0\ K)]_{St}$$
$$- c \cdot [H^\theta(N, 298\ K) - H^\theta(N, 0\ K)]_{St}$$
$$- d \cdot [H^\theta(O, 298\ K) - H^\theta(O, 0\ K)]_{St}, \tag{B.1}$$

whereby the terms in the square brackets are the heat capacity corrections and indicate the enthalpy changes due to raising the temperature from 0 K to 298 K. The $[H^\theta(298\ K) - H^\theta(0\ K)]$ corrections for $C_aH_bN_cO_d$ molecules can be extracted from the output of Gaussian program package [463] and were taken from CRC Handbook [464] for the standard states of elements.

The standard enthalpies of formation at 0 K were calculated by subtracting the calculated values of the atomization energies of $C_aH_bN_cO_d$ compounds from the experimental enthalpies of formation of the isolated atoms (equation (B.2)) [273, 462].

$$\Delta_f H^\theta(C_aH_bN_cO_d, 0\ K) = a \cdot \Delta_f H^\theta(C, 0\ K)$$
$$+ b \cdot \Delta_f H^\theta(H, 0\ K)$$
$$+ c \cdot \Delta_f H^\theta(N, 0\ K)$$
$$+ d \cdot \Delta_f H^\theta(O, 0\ K)$$
$$- \Delta_a H^\theta(C_aH_bN_cO_d, 0\ K). \tag{B.2}$$

The $\Delta_f H^\theta(atom, 0\ K)$ values, i.e. the atomic enthalpies of formation of the elements in their standard states at 0 K, were obtained from standard thermodynamic tables [465, 466]. The $\Delta_a H^\theta(C_aH_bN_cO_d, 0\ K)$ values, i.e. the atomization (dissociation) energies of $C_aH_bN_cO_d$ compounds, were obtained by subtracting the quantum mechanical energy of the molecule (electronic energy + zero-point energy) from the quantum mechanical energies (electronic energies) of the atoms (equation (B.3)) [273, 462]:

$$\Delta_a H^\theta(C_aH_bN_cO_d, 0\ K) = a \cdot E_e(C) + b \cdot E_e(H) + c \cdot E_e(N) + d \cdot E_e(O)$$
$$- E_0(C_aH_bN_cO_d). \tag{B.3}$$

The $E_e(atom)$ and $E_0(C_aH_bN_cO_d)$ values were extracted from the output of Gaussian program package [463].

https://doi.org/10.1515/9783110740158-016

Table B.1: The B3LYP/6-31G* calculated total energies and formation enthalpies (at 0 K) for 100 energetic materials with high nitrogen contents.

Formula	Name	E_e (Hartrees)[a]	E_0 (Hartrees)[b]	$H_{298}^{\theta} - H_0^{\theta}$ (kJ/mol)[c]	$\Delta_f H^{\theta}$ (g) (0 K) (kJ/mol)[d]
CHN_5O_2	5-Nitro-1H-tetrazole	−462.73092	−462.68250	20.3	381.2
CHN_7	5-Azido-1H-tetrazole	−421.83034	−421.78146	20.9	694.1
CH_2N_2	Cyanamide	−148.78006	−148.74651	13.6	166.1
CH_2N_4	1H-Tetrazole	−258.25090	−258.20483	14.0	347.6
CH_2N_4O	1H-Tetrazol-5-ol	−333.47117	−333.42062	16.8	187.1
$CH_2N_6O_2$	Nitroguanylazide	−518.07074	−518.00826	27.4	435.0
$CH_2N_6O_2$	5-Nitroaminotetrazole	−518.08423	−518.01890	24.1	407.1
$CH_2N_6O_2$	5-Nitriminotetrazole	−518.08498	−518.01950	23.7	405.5
CH_3N_5	1H-Tetrazol-5-amine	−313.60883	−313.54613	18.5	360.7
CH_3N_5	1H-Tetrazol-1-amine	−313.57178	−313.50903	18.3	458.1
$CH_4N_4O_2$	1-Nitroguanidine	−409.83771	−409.76022	23.9	134.9
CH_4N_6	1H-Tetrazol-1,5-diamine	−368.93317	−368.85361	22.5	462.5
CH_5N_3	Guanidine	−205.36258	−205.28765	18.0	88.0
$CH_5N_5O_2$	N-Nitro-N'-aminoguanidine	−465.15588	−465.06137	27.8	253.4
$C_2N_6O_3$	5,6-(3,4-Furazano)-1,2,3,4-tetrazine-1,3-dioxide	−630.05517	−630.00396	24.0	726.6
C_2N_{10}	3,6-Diazido-1,2,4,5-tetrazine	−623.50507	−623.44884	29.7	1110.1
C_2HN_5	1H-Tetrazole-5-carbonitrile	−350.48131	−350.43683	18.2	521.9
$C_2H_2N_6$	3-Azido-1H-1,2,4-triazole	−405.83770	−405.77581	21.6	541.1
$C_2H_2N_8$	1H,1'H-5,5'-Bitetrazole	−515.31771	−515.24441	24.2	695.6
$C_2H_3N_3$	1H-1,2,3-Triazole	−242.22232	−242.16432	14.9	286.1
$C_2H_3N_3$	1H-1,2,4-Triazole	−242.24927	−242.19042	14.9	217.6
$C_2H_3N_5O_2$	5-Amino-3-nitro-1H-1,2,4-triazole	−502.09665	−502.01906	26.0	238.8
$C_2H_3N_5O_2$	5-Nitro-1H-1,2,4-triazol-3-amine	−502.09856	−502.02088	25.7	234.0
$C_2H_3N_5O_2$	N-Nitro-1H-1,2,4-triazol-3-amine	−502.08080	−502.00273	25.2	281.7
$C_2H_3N_7$	3-Azido-4H-1,2,4-triazol-4-amine	−461.14442	−461.06652	26.3	687.0
$C_2H_3N_9$	di(1H-Tetrazol-5-yl)amine	−570.66899	−570.57935	28.5	725.3

Table B.1 (continued)

Formula	Name	E_e (Hartrees)[a]	E_0 (Hartrees)[b]	$H^{\theta}_{298} - H^{\theta}_0$ (kJ/mol)[c]	$\Delta_f H^{\theta}$ (g) (0 K) (kJ/mol)[d]
$C_2H_4N_4$	1H-1,2,4-Triazole-3-amine	-297.60778	-297.53239	19.2	228.9
$C_2H_4N_4$	4H-1,2,4-Triazol-4-amine	-297.55301	-297.47836	19.2	370.7
$C_2H_4N_4$	1-Methyl-1H-tetrazole	-297.56746	-297.49387	20.0	330.0
$C_2H_4N_4$	5-Methyl-1H-tetrazole	-297.57591	-297.50238	19.8	307.7
$C_2H_4N_4$	2-Methyl-1H-tetrazole	-297.57441	-297.50014	19.5	313.6
$C_2H_4N_4$	2-Methyl-2H-tetrazole	-297.59010	-297.51710	22.3	269.0
$C_2H_4N_4$	Dicyandiamide	-297.77902	-297.70045	22.4	193.7
$C_2H_4N_4O$	5-Methoxy-1H-tetrazole	-372.75583	-372.67773	23.4	253.3
$C_2H_4N_4O$	3,4-Diaminofurazan	-372.77902	-372.70045	24.4	453.8
$C_2H_4N_6$	1,2,4,5-tetrazine-3,6-diamine	-407.05804	-406.97417	29.1	378.5
$C_2H_4N_6O_2$	3,6-Diamino-1,2,4,5-tetrazine 1,4-dioxide	-557.40466	-557.31215	29.6	381.2
$C_2H_4N_6O_2$	1-Methyl-5-nitriminotetrazole	-557.40420	-557.31112	30.1	387.9
$C_2H_4N_6O_2$	2-Methyl-5-nitraminotetrazole	-557.40175	-557.30855	37.1	709.9
$C_2H_4N_8O_2$	1,3-Diazido-2-nitro-2-azapropane	-666.81480	-666.71334	32.7	876.4
$C_2H_4N_{10}$	5,5'-Hydrazotetrazole	-625.97430	-625.86810	24.1	343.1
$C_2H_5N_5$	1-Methyl-1H-tetrazol-5-amine	-352.92560	-352.83517	23.8	325.3
$C_2H_5N_5$	2-Methyl-1H-tetrazol-5-amine	-352.93288	-352.84193	23.8	325.3
$C_2H_5N_5$	2-Methyl-2H-tetrazol-5-amine	-352.93288	-352.84193	23.9	363.6
$C_2H_5N_5$	5-Methylamino-1H-tetrazole	-352.91787	-352.82735	27.7	391.6
$C_2H_5N_7$	5-Guanylaminotetrazole	-462.44641	-462.34413	33.7	754.1
$C_2H_6N_8$	3,6-Dihydrazino-1,2,4,5-tetrazine	-517.66931	-517.55236	37.1	1129.5
C_3N_{12}	2,4,6-Triazido-1,3,5-triazine	-771.15932	-771.08616	41.4	880.4
$C_3HN_{11}O_2$	6-Nitroamino-2,4-diazido[1,3,5]triazine	-867.39800	-867.30913	20.2	650.5
$C_3H_2N_6$	Tetrazolo[1,5-b][1,2,4]triazine	-443.91956	-443.85135	29.7	561.1
$C_3H_2N_6O_2$	2-(5-Nitro-1H-tetrazol-1-yl)acetonitrile	-594.26953	-594.19472	23.9	480.0
$C_3H_3N_5$	2-Methyl-2H-tetrazole-5-carbonitrile	-389.80767	-389.73514	35.8	729.7
$C_3H_3N_9O_2$	5-((5-Nitro-2H-tetrazol-2-yl)methyl)-1H-tetrazole	-759.10801	-759.00420		

Table B.1 (continued)

Formula	Name	E_e (Hartrees)[a]	E_0 (Hartrees)[b]	$H_{298}^\theta - H_0^\theta$ (kJ/mol)[c]	$\Delta_f H^\theta$ (g) (0 K) (kJ/mol)[d]
$C_3H_4N_6$	5,8-Dihydrotetrazolo[1,5-b][1,2,4]triazine	-445.10641	-445.01526	24.2	653.7
$C_3H_4N_6$	3-Azido-5-methyl-1H-1,2,4-triazole	-445.16235	-445.07337	28.1	501.1
$C_3H_4N_6O_2$	4-Nitro-5,6-dihydro-4H-imidazo[1,2-d]tetrazole	-595.48066	-595.38046	29.3	506.9
$C_3H_4N_8O_4$	5,8-Dinitro-5,6,7,8-tetrahydrotetrazolo[1,5-b][1,2,4]triazine	-855.27837	-855.16010	39.2	656.8
$C_3H_4N_8O_4$	1-Nitroguanyl-3-nitro-5-amino-1,2,4-triazole	-855.37877	-855.26092	42.5	392.1
$C_3H_5N_5O$	5-Acetamidotetrazole	-466.26538	-466.16560	28.9	189.3
$C_3H_5N_5O$	4,6-Diamino-1,3,5-triazin-2(1H)-one	-466.35186	-466.24972	28.5	-31.5
$C_3H_5N_7$	3-Azido-5-methyl-4H-1,2,4-triazol-4-amine	-500.46950	-500.36435	32.5	646.3
$C_3H_6N_4$	1,5-Dimethyl-1H-tetrazole	-336.89210	-336.79098	25.4	291.2
$C_3H_6N_4$	2,5-Dimethyl-2H-tetrazole	-336.89818	-336.79677	25.6	276.1
$C_3H_6N_6$	1,3,5-Triazine-2,4,6-triamine	-446.49172	-446.37702	29.5	137.4
$C_3H_6N_6$	5,6,7,8-Tetrahydrotetrazolo[1,5-b][1,2,4]triazine	-446.33547	-446.22037	26.5	548.7
$C_3H_6N_6O_2$	1,6-Dimethyl-5-nitraminotetrazole	-596.71121	-596.59053	35.2	388.9
$C_3H_6N_8$	1-(2-Azidoethyl)-1H-tetrazol-5-amine	-555.81867	-555.69632	35.5	683.9
$C_3H_6N_{10}O_4$	1,5-Diazido-2,4-dinitro-2,4-diazapentane	-965.95284	-965.80396	52.5	784.6
$C_3H_7N_5$	1-Methyl-5-methylaminotetrazole	-392.23449	-392.11628	29.6	346.3
$C_3H_7N_5$	5-(Dimethylamino)-tetrazole	-392.22789	-392.10972	29.1	363.6
$C_3H_7N_5O$	2-(5-Amino-1H-tetrazol-1-yl)ethan-1-ol	-467.44441	-467.32120	32.0	214.3
$C_4H_4N_{10}O_5$	5,5'-Dinitro-3,3'-azoxy-1,2,4-triazole	-1076.90170	-1076.78507	47.5	681.0
$C_4H_4N_8$	4,4'-Azobis-1,2,4-triazole	-592.68104	-592.57609	31.4	873.8
$C_4H_4N_8O_2$	3,3'-Diamino-4,4'-azofurazan	-743.06077	-742.94723	38.2	711.4
$C_4H_4N_8O_3$	3,3'-Diamino-4,4'-azoxyfurazan	-818.23777	-818.11919	40.4	665.9
$C_4H_4N_{12}$	6,6'-(Diazene-1,2-diyl)bis(1,2,4,5-tetrazin-3-amine)	-811.63562	-811.51265	43.5	1184.5
$C_4H_4N_{14}$	3,6-Bis(1H-1,2,3,4-tetrazol-5-ylamino)-1,2,4,5-tetrazine	-921.17726	-921.03922	44.7	1186.8
$C_4H_6N_4$	2-Methyl-5-vinyl-2H-tetrazole	-374.97864	-374.87194	27.2	386.5
$C_4H_6N_6$	3-Azido-5-ethyl-4H-1,2,4-triazole	-484.45817	-484.34170	32.8	537.9

Table B.1 (continued)

Formula	Name	E_e (Hartrees)[a]	E_0 (Hartrees)[b]	$H_{298}^\theta - H_0^\theta$ (kJ/mol)[c]	$\Delta_f H^\theta$(g) (0 K) (kJ/mol)[d]
$C_4H_6N_6$	3-(5-Amino-1H-tetrazol-1-yl) propanenitrile	−484.47524	−484.35765	32.9	496.1
$C_4H_6N_6$	3-(5-Amino-2H-tetrazol-2-yl) propanenitrile	−484.48380	−484.36570	32.6	474.9
$C_4H_6N_8$	1,2-Di(1H-tetrazol-5-yl)ethane	−593.94862	−593.81921	34.8	669.0
$C_4H_6N_8O_2$	4,4'-Hydrazobis-(1,2,5-oxadiazol-3-amine)	−744.28136	−744.14276	40.3	631.6
$C_4H_6N_{10}$	(cis)1,1'-Dimethyl-5,5'-azotetrazole	−703.38229	−703.24518	41.3	935.4
$C_4H_6N_{10}$	(trans)1,1'-Dimethyl-5,5'-azotetrazole	−703.38843	−703.25123	41.6	919.5
$C_4H_6N_{10}$	2,2'-Dimethyl-5,5'-azotetrazole	−703.39638	−703.25871	41.5	899.9
$C_4H_6N_{12}O_4$	3,6-Bis(nitroguanyl)-1,2,4,5-tetrazine	−1113.66254	−1113.49599	56.3	660.3
$C_4H_7N_5$	5-Amino-1-(2-propenyl)-1H-tetrazole	−430.31587	−430.19265	31.2	453.6
$C_4H_7N_5$	5-Amino-2-(2-propenyl)-2H-tetrazole	−430.32384	−430.20011	30.9	434.0
$C_4H_7N_9$	bis(2-Methyl-2H-tetrazol-5-yl)amine	−649.31166	−649.16587	40.0	668.0
$C_4H_8N_8O_2$	1,5-Diazido-3-nitro-3-azapentane	−745.44366	−745.28588	47.8	689.3
$C_4H_8N_{10}$	3,6-Diguanidino-1,2,4,5-tetrazine	−704.68475	−704.52319	44.5	639.0
$C_4H_8N_{10}O_4$	1,6-Diazido-2,5-dinitro-2,5-diazahexane	−1005.26729	−1005.09000	57.5	774.9
$C_4H_8N_{12}O_6$	1,7-Diazido-2,4,6-trinitro-2,4,6-triazaheptane	−1265.08978	−1264.89317	67.0	863.0
$C_5H_9N_9$	N,2-Dimethyl-N-(2-methyl-2H-tetrazol-5-yl)-2H-tetrazol-5-amine	−688.61739	−688.44403	45.8	679.0
C_6N_{20}	4,4',6,6'-Tetra(azido)azo-1,3,5-triazine	−1323.37055	−1323.24159	63.9	1933.0
$C_6H_2N_{20}$	4,4',6,6'-Tetra(azido)hydrazo-1,3,5-triazine	−1324.63571	−1324.48119	65.9	1737.4
$C_6H_{10}N_{10}$	2,2'-Diethyl-5,5'-azotetrazole	−782.03174	−781.83751	50.7	862.9
$C_6H_{14}N_{16}$	1,2-bis(4,6-Dihydrazinyl-1,3,5-triazin-2-yl)hydrazine	−1112.97843	−1112.70251	74.8	987.7
$C_7H_7N_9$	3-Azido-6-(3,5-dimethylpyrazol-1-yl)-1,2,4,5-tetrazine	−763.55913	−763.40099	46.9	974.1
$C_8H_4N_{16}O_6$	Bis[4-aminofurazanyl-3-azoxy]azofurazan	−1634.00217	−1633.80967	75.4	1590.5

[a] Electronic energy at 0 K. The E_e energies for atoms (in Hartrees) were calculated as: C(−37.846280), H(−0.500273), N(−54.584489), O(−75.060623).
[b] Molecular energy with zero-point energy correction.
[c] Enthalpy difference from 298 K to 0 K. The $[H^\theta(298\,K) - H^\theta(0\,K)]_{st}$ for atoms (in kJ/mol) were obtained as: C(1.050), H(4.234), N(4.335), O(4.340) [464].
[d] Enthalpy of formation at 0 K. The $\Delta_f H^\theta$ for atoms in their standard states at 0 K (in kJ/mol) were obtained as: C(711.38), H(216.03), N(470.57), O(246.84) [465, 466].

Table B.2: Predicted gas phase standard enthalpies of formation for 100 energetic materials with high nitrogen content.

Formula	Name	Predicted $\Delta_f H^\theta$ (g)(298 K) (kJ/mol)	
		B3LYP/6-31G*	**PM6**
CHN_5O_2	5-Nitro-1H-tetrazole	365.8	386.2
CHN_7	5-Azido-1H-tetrazole	679.4	672.6
CH_2N_2	Cyanamide	161.5	147.7
CH_2N_4	1H-Tetrazole	334.7	336.3
CH_2N_4O	1H-Tetrazol-5-ol	172.7	177.2
$CH_2N_6O_2$	Nitroguanylazide	418.2	365.8
$CH_2N_6O_2$	5-Nitroaminotetrazole	387.0	367.6
$CH_2N_6O_2$	5-Nitriminotetrazole	385.0	366.7
CH_3N_5	1H-Tetrazol-5-amine	343.7	345.0
CH_3N_5	1H-Tetrazol-1-amine	440.9	424.8
$CH_4N_4O_2$	1-Nitroguanidine	114.8	77.0
CH_4N_6	1H-Tetrazol-1,5-diamine	441.1	430.7
CH_5N_3	Guanidine	70.8	68.8
$CH_5N_5O_2$	N-Nitro-N'-aminoguanidine	228.7	176.3
$C_2N_6O_3$	5,6-(3,4-Furazano)-1,2,3,4-tetrazine-1,3-dioxide	709.5	682.7
C_2N_{10}	3,6-Diazido-1,2,4,5-tetrazine	1094.4	1095.2
C_2HN_5	1H-Tetrazole-5-carbonitrile	512.1	510.4
$C_2H_2N_6$	3-Azido-1H-1,2,4-triazole	526.1	569.3
$C_2H_2N_8$	1H,1'H-5,5'-Bitetrazole	674.6	688.8
$C_2H_3N_3$	1H-1,2,3-Triazole	273.2	251.2
$C_2H_3N_3$	1H-1,2,4-Triazole	204.6	220.2
$C_2H_3N_5O_2$	5-Amino-3-nitro-1H-1,2,4-triazole	219.7	264.3
$C_2H_3N_5O_2$	5-Nitro-1H-1,2,4-triazol-3-amine	214.5	266.5
$C_2H_3N_5O_2$	N-Nitro-1H-1,2,4-triazol-3-amine	261.7	257.2
$C_2H_3N_7$	3-Azido-4H-1,2,4-triazol-4-amine	668.1	644.4
$C_2H_3N_9$	di(1H-Tetrazol-5-yl)amine	700.0	727.7
$C_2H_4N_4$	1H-1,2,4-Triazole-3-amine	211.7	229.9
$C_2H_4N_4$	4H-1,2,4-Triazol-4-amine	353.6	316.5
$C_2H_4N_4$	1-Methyl-1H-tetrazole	313.6	326.5
$C_2H_4N_4$	5-Methyl-1H-tetrazole	291.1	289.4
$C_2H_4N_4$	2-Methyl-2H-tetrazole	296.7	325.4
$C_2H_4N_4$	Dicyandiamide	254.9	244.7
$C_2H_4N_4O$	5-Methoxy-1H-tetrazole	175.3	185.8
$C_2H_4N_4O$	3,4-Diaminofurazan	236.0	238.4
$C_2H_4N_6$	1,2,4,5-tetrazine-3,6-diamine	433.2	398.9
$C_2H_4N_6O_2$	3,6-Diamino-1,2,4,5-tetrazine 1,4-dioxide	353.8	391.9
$C_2H_4N_6O_2$	1-Methyl-5-nitriminotetrazole	357.1	356.5
$C_2H_4N_6O_2$	2-Methyl-5-nitraminotetrazole	364.2	374.0
$C_2H_4N_8O_2$	1,3-Diazido-2-nitro-2-azapropane	684.6	620.3
$C_2H_4N_{10}$	5,5'-Hydrazotetrazole	846.7	814.6
$C_2H_5N_5$	1-Methyl-1H-tetrazol-5-amine	322.3	337.2
$C_2H_5N_5$	2-Methyl-1H-tetrazol-5-amine	304.2	338.1

Table B.2 (continued)

Formula	Name	Predicted $\Delta_f H^\theta$ (g)(298 K) (kJ/mol)	
		B3LYP/6-31G*	PM6
$C_2H_5N_5$	2-Methyl-2H-tetrazol-5-amine	304.2	338.1
$C_2H_5N_5$	5-Methylamino-1H-tetrazole	342.6	342.0
$C_2H_5N_7$	5-Guanylaminotetrazole	365.7	387.6
$C_2H_6N_8$	3,6-Dihydrazino-1,2,4,5-tetrazine	725.6	564.8
C_3N_{12}	2,4,6-Triazido-1,3,5-triazine	1111.4	1236.2
$C_3HN_{11}O_2$	6-Nitroamino-2,4-diazido[1,3,5]triazine	858.1	933.3
$C_3H_2N_6$	Tetrazolo[1,5-b][1,2,4]triazine	633.1	661.0
$C_3H_2N_6O_2$	2-(5-Nitro-1H-tetrazol-1-yl)acetonitrile	544.5	542.2
$C_3H_3N_5$	2-Methyl-2H-tetrazole-5-carbonitrile	466.4	494.5
$C_3H_3N_9O_2$	5-((5-Nitro-2H-tetrazol-2-yl)methyl)-1H-tetrazole	702.0	738.1
$C_3H_4N_6$	5,8-Dihydrotetrazolo[1,5-b][1,2,4]triazine	631.9	561.6
$C_3H_4N_6$	3-Azido-5-methyl-1H-1,2,4-triazole	483.2	521.8
$C_3H_4N_6O_2$	4-Nitro-5,6-dihydro-4H-imidazo[1,2-d]tetrazole	481.4	457.5
$C_3H_4N_8O_4$	5,8-Dinitro-5,6,7,8-tetrahydrotetrazolo[1,5-b][1,2,4]triazine	623.8	568.4
$C_3H_4N_8O_4$	1-Nitroguanyl-3-nitro-5-amino-1,2,4-triazole	362.5	355.1
$C_3H_5N_5O$	5-Acetamidotetrazole	167.9	140.3
$C_3H_5N_5O$	4,6-Diamino-1,3,5-triazin-2(1H)-one	−53.3	−35.7
$C_3H_5N_7$	3-Azido-5-methyl-4H-1,2,4-triazol-4-amine	624.1	601.4
$C_3H_6N_4$	1,5-Dimethyl-1H-tetrazole	270.8	282.4
$C_3H_6N_4$	2,5-Dimethyl-2H-tetrazole	255.8	284.4
$C_3H_6N_6$	1,3,5-Triazine-2,4,6-triamine	112.3	177.8
$C_3H_6N_6$	5,6,7,8-Tetrahydrotetrazolo[1,5-b][1,2,4]triazine	520.6	473.8
$C_3H_6N_6O_2$	1,6-Dimethyl-5-nitraminotetrazole	360.8	355.5
$C_3H_6N_8$	1-(2-Azidoethyl)-1H-tetrazol-5-amine	656.1	645.9
$C_3H_6N_{10}O_4$	1,5-Diazido-2,4-dinitro-2,4-diazapentane	747.9	642.3
$C_3H_7N_5$	1-Methyl-5-methylaminotetrazole	321.4	335.1
$C_3H_7N_5$	5-(Dimethylamino)-tetrazole	338.2	337.3
$C_3H_7N_5O$	2-(5-Amino-1H-tetrazol-1-yl)ethan-1-ol	187.5	132.2
$C_4H_2N_{10}O_5$	5,5′-Dinitro-3,3′−azoxy-1,2,4-triazole	650.8	760.0
$C_4H_4N_8$	4,4′-Azobis-1,2,4-triazole	849.4	800.7
$C_4H_4N_8O_2$	3,3′−Diamino-4,4′-azofurazan	685.1	730.3
$C_4H_4N_8O_3$	3,3′−Diamino-4,4′-azoxyfurazan	637.5	676.2
$C_4H_4N_{12}$	6,6′-(Diazene-1,2-diyl)bis(1,2,4,5-tetrazin-3-amine)	1154.9	1075.4
$C_4H_4N_{14}$	3,6-Bis(1H-1,2,3,4-tetrazol-5-ylamino)-1,2,4,5-tetrazine	1149.7	1164.4
$C_4H_6N_4$	2-Methyl-5-vinyl-2H-tetrazole	366.7	386.6
$C_4H_6N_6$	3-Azido-5-ethyl-4H-1,2,4-triazole	515.1	488.9
$C_4H_6N_6$	3-(5-Amino-1H-tetrazol-1-yl) propanenitrile	473.3	456.0
$C_4H_6N_6$	3-(5-Amino-2H-tetrazol-2-yl) propanenitrile	451.9	456.2
$C_4H_6N_8$	1,2-Di(1H-tetrazol-5-yl)ethane	639.5	613.9
$C_4H_6N_8O_2$	4,4′-Hydrazobis-(1,2,5-oxadiazol-3-amine)	598.9	593.8
$C_4H_6N_{10}$	(cis)1,1′-Dimethyl-5,5′-azotetrazole	903.8	927.9
$C_4H_6N_{10}$	(trans)1,1′-Dimethyl-5,5′-azotetrazole	888.2	929.1

Table B.2 (continued)

Formula	Name	Predicted $\Delta_f H^\theta$(g)(298 K) (kJ/mol)	
		B3LYP/6-31G*	PM6
$C_4H_6N_{10}$	2,2'-Dimethyl-5,5'-azotetrazole	868.4	948.9
$C_4H_6N_{12}O_4$	3,6-Bis(nitroguanyl)-1,2,4,5-tetrazine	617.6	572.9
$C_4H_7N_5$	5-Amino-1-(2-propenyl)-1H-tetrazole	429.3	403.5
$C_4H_7N_5$	5-Amino-2-(2-propenyl)-2H-tetrazole	409.4	409.3
$C_4H_7N_9$	bis(2-Methyl-2H-tetrazol-5-yl)amine	635.2	722.3
$C_4H_8N_8O_2$	1,5-Diazido-3-nitro-3-azapentane	655.7	584.2
$C_4H_8N_{10}$	3,6-Diguanidino-1,2,4,5-tetrazine	602.1	604.3
$C_4H_8N_{10}O_4$	1,6-Diazido-2,5-dinitro-2,5-diazahexane	733.6	626.7
$C_4H_8N_{12}O_6$	1,7-Diazido-2,4,6-trinitro-2,4,6-triazaheptane	813.9	675.7
$C_5H_9N_9$	N,2-Dimethyl-N-(2-methyl-2H-tetrazol-5-yl)-2H-tetrazol-5-amine	642.4	724.6
C_6N_{20}	4,4',6,6'-Tetra(azido)azo-1,3,5-triazine	1903.9	2102.8
$C_6H_2N_{20}$	4,4',6,6'-Tetra(azido)hydrazo-1,3,5-triazine	1701.9	1898.1
$C_6H_{10}N_{10}$	2,2'-Diethyl-5,5'-azotetrazole	821.6	884.9
$C_6H_{14}N_{16}$	1,2-bis(4,6-Dihydrazinyl-1,3,5-triazin-2-yl)hydrazine	927.6	816.2
$C_7H_7N_9$	3-Azido-6-(3,5-dimethylpyrazol-1-yl)-1,2,4,5-tetrazine	944.9	914.3
$C_8H_4N_{16}O_6$	Bis[4-aminofurazanyl-3-azoxy]azofurazan	1545.1	1625.4

C Glossary of compound names as well as the measured and calculated values of the condensed phase heats of formation for some energetic ionic liquids and salts

Table C.1: Predictions of $\Delta_f H^\theta(c)[IMILoS]$ for some different energetic imidazolium based ionic liquids or salts as compared to experimental data and the predicted quantum calculated values based on Byrd and Rice method [467].

Name		$\Delta_f H^\theta(c)[IMILoS]$ (kJ/mol)		
Cation	Anion	Exp. or Pred.[a]	Equation (2.18)	Dev
1-Methyl-3-octylimidazolium	Di-1H-tetrazol-5-ylazanide	539.6 [169]	533.3	−6.3
1-Hexyl-3-methylimidazolium	Di-1H-tetrazol-5-ylazanide	615.9 [169]	607.1	−8.8
1-Butyl-3-methylimidazolium	Di-1H-tetrazol-5-ylazanide	682.3 [169]	680.9	−1.4
1-Ethyl-3-methylimidazolium	Di-1H-tetrazol-5-ylazanide	725.6 [169]	754.8	29.2
1,2,3-Trimethylimidazolium	Di-1H-tetrazol-5-ylazanide	652.3 [169]	663.4	11.1
1,2,3-Trimethylimidazolium	Di-1H-tetrazol-5-ylazanide	774.7 [169]	783.2	8.5
1,3-Dimethylimidazolium	Di-1H-tetrazol-5-ylazanide	752.3 [169]	816.1	63.8
1-Ethyl-3-methylimidazolium	[(Cyanoimino)methylene]azanide	235.3 [468]	222.2	−13.1
1-Butyl-3-methylimidazolium	[(Cyanoimino)methylene]azanide	206.2 [469]	221.2	15.0
1-Ethyl-3-methylimidazolium	Ethyl sulfate	−579.1 [470]	−584.3	−5.2
1-Butyl-3-methylimidazolium	(E)-5,5'-(diazene-1,2-diyl)bis(tetrazol-1-ide)	−1006 [471]	−938.9	67.1
1,3-Dimethyl-5-nitro-1H-imidazol-3-ium	(E)-5,5'-(diazene-1,2-diyl)bis(tetrazol-1-ide)	1345 [471]	1241.2	−103.8
1-Ethyl-3-methylimidazolium	Bis[(trifluoromethyl)sulfonyl]azanide	−1917.2	−1843.2	74.0
1-Hexyl-3-methylimidazolium	Bis[(trifluoromethyl)sulfonyl]azanide	−2006.4	−2019.8	−13.4
1-Methyl-3-octylimidazolium	Bis[(trifluoromethyl)sulfonyl]azanide	−2052.0	−2035.7	16.3
1-Decyl-3-methyl-1H-imidazol-3-ium	Bis[(trifluoromethyl)sulfonyl]azanide	−2095.7	−2051.7	44.0
1-Tetradecyl-3-methylimidazolium	Bis[(trifluoromethyl)sulfonyl]azanide	−4871.5	−4892.4	−20.9
1-Butyl-3-methylimidazolium	Hexafluorophosphate	−2105.3	−2148.5	−43.2
1-Hexyl-3-methylimidazolium	Hexafluorophosphate	−2144.3	−2164.4	−20.1
1-Methyl-3-octylimidazolium	Hexafluorophosphate	−2186.2	−2180.4	5.8
1-Decyl-3-methyl-1H-imidazol-3-ium	Hexafluorophosphate	−2226.1	−2196.3	29.8
1-Dodecyl-3-methylimidazolium	Hexafluorophosphate	−4693.4	−4678.4	15.0

https://doi.org/10.1515/9783110740158-017

Table C.1 (continued)

Name		$\Delta_f H^\theta$(c)[IMILoS] (kJ/mol)		
Cation	Anion	Exp. or Pred.[a]	Equation (2.18)	Dev
1-Butyl-3-methylimidazolium	Tetrafluoroborate	−1662.7	−1647.7	15.0
1-Methyl-3-octylimidazolium	Tetrafluoroborate	−1740.3	−1893.1	−152.8
1-Dodecyl-3-methylimidazolium	Tetrafluoroborate	−4245.3	−4238.5	6.8
1-Butyl-2,3-dimethylimidazolium	Bromide	−230.2	−220.6	9.6
1-Pentyl-2,3-dimethylimidazolium	Bromide	−247.0	251.5	498.5
1-Octyl-2,3-dimethylimidazolium	Bromide	−305.3	−313.8	−8.5
1,3-Dibutyl-1H-imidazol-3-ium	Bromide	−248.2	−252.0	−3.8
1,3-Pentyl-1H-imidazol-3-ium	Bromide	−289.6	−313.8	−24.2
1,3-Octyl-1H-imidazol-3-ium	Bromide	−322.8	−303.8	19.0

[a]In the cases which the experimental values were not available, the predicted quantum calculated values based on Byrd and Rice method [467] were used.

Table C.2: Predictions of $\Delta_f H^\theta$(c)[TAILoS] for some different energetic triazolium based ionic liquids or salts as compared to experimental data.

Cation	Anion	Exp.	Equation (2.18)	Dev.
3-Azido-3H-1,2,4-triazol-4-ium	Nitrate	218.8 [472]	201.9	−16.9
3-Azido-1H-1,2,4-triazol-4-ium	5-Nitrotetrazol-1-ide	979.9 [472]	968.3	−11.6
3-azido-3H-1,2,4-triazol-4-ium	4,5-Dinitroimidazol-1-ide	401.7 [472]	411.5	9.8
1H-1,2,4-triazol-4-ium	5-Nitrotetrazol-1-ide	409.6 [472]	433.5	23.9
1H-1,2,4-triazol-4-ium	4,5-Dinitroimidazol-1-ide	503.3 [472]	456.5	−46.8
1H-1,2,4-triazol-4-ium	2,4,6-Ttrinitrophenolate	497.9 [472]	493.8	−4.1
1-Amino-4H-1,2,4-triazol-1-ium	Nitrate	34.7 [472]	−58.1	−92.8
1-Amino-4H-1,2,4-triazol-1-ium	Perchlorate	356.9 [472]	403.3	46.4
1-Amino-4H-1,2,4-triazol-1-ium	Nitrate	−109.6 [472]	−58.1	51.5
1-Amino-4H-1,2,4-triazol-1-ium	Perchlorate	298.3 [472]	291.5	−6.8
1-Amino-4H-1,2,4-triazol-1-ium	5-Nitrotetrazol-1-ide	702.1 [472]	708.3	6.2
1-amino-4H-1,2,4-triazol-1-ium	4-Nitro-1,2,3-triazol-1-ide	141.4 [472]	143.4	1.9
1-Amino-4H-1,2,4-triazol-1-ium	4,5-Dinitroimidazol-1-ide	466.5 [472]	412.2	−54.3
1-Amino-4H-1,2,4-triazol-1-ium	2,4,6-Trinitrophenolate	469.0 [472]	449.5	−19.5
3-Amino-1H-1,2,4-triazol-1-ium	Nitrate	−171.1 [473]	−169.9	1.2
1,5-Diamino-4H-1,2,4-triazol-1-ium	Nitrate	−89.5 [472]	−94.5	−5.0
1,5-Diamino-4H-1,2,4-triazol-1-ium	Perchlorate	484.5 [472]	478.7	−5.8
1,5-Diamino-4H-1,2,4-triazol-1-ium	4,5-Dinitroimidazol-1-ide	405.8 [472]	487.6	81.7
3,4,5-Triamino-4H-1,2,4-triazol-1-ium	5-Nitrotetrazol-1-ide	528.0 [474]	510.6	−17.4

Table C.2 (continued)

Cation	Anion	Exp.	Equation (2.18)	Dev.
3-Azido-1-methyl-4H-1,2,4-triazol-1-ium	Nitrate	93.3 [472]	147.3	54.0
3-Azido-1-methyl-4H-1,2,4-triazol-1-ium	Perchlorate	574.9 [472]	608.7	33.8
3-Azido-1-methyl-1H-1,2,4-triazol-4-ium	5-Nitrotetrazol-1-ide	1085.3 [472]	1113.1	27.8
3-Azido-1-methyl-4H-1,2,4-triazol-1-ium	4,5-Dinitroimidazol-1-ide	700.8 [472]	617.6	−83.2
3-Azido-5-methyl-4H-1,2,4-triazol-1-ium	Nitrate	156.5 [472]	147.3	−9.2
1-Methyl-1H-1,2,4-triazol-4-ium	5-Nitrotetrazol-1-ide	808.3 [472]	777.7	−30.6
1-Methyl-3H-1,2,4-triazol-1-ium	4,5-Dinitroimidazol-1-ide	80.3 [472]	95.9	15.6
1-Methyl-3H-1,2,4-triazol-1-ium	2,4,6-Trinitrophenolate	409.6 [472]	439.2	29.6
1-Amino-4-methyl-4H-1,2,4-triazol-1-ium	Nitrate	−160.2 [472]	−112.8	47.5
1-Amino-4-methyl-4H-1,2,4-triazol-1-ium	Perchlorate	544.8 [472]	548.1	3.3
4-Amino-1-methyl-4H-1,2,4-triazol-1-ium	Nitrate	−172.8 [472]	−112.8	60.0
4-Amino-1-methyl-4H-1,2,4-triazol-1-ium	Perchlorate	215.1 [472]	236.9	21.8
4-Amino-1-methyl-4H-1,2,4-triazol-1-ium	3,5-Dinitro-1,2,4-triazol-1-ide	378.2 [472]	350.5	−27.7
3,4,5-Triamino-1-methyl-4H-1,2,4-triazol-1-ium	Azide	530.6 [475]	506.3	−24.3
3,4,5-Triamino-1-methyl-4H-1,2,4-triazol-1-ium	Nitrate	64.2 [475]	38.0	−26.2
3,4,5-Triamino-1-methyl-4H-1,2,4-triazol-1-ium	Perchlorate	498.4 [475]	499.4	1.0
3,4,5-Triamino-1-methyl-4H-1,2,4-triazol-1-ium	Dinitroamide	276.3 [475]	242.1	−34.2
3,4,5-Triamino-1-methyl-4H-1,2,4-triazol-1-ium	5-Nitrotetrazol-1-ide	519.0 [475]	605.0	86.0
3-Azido-1-(2-azidoethyl)-1H-1,2,4-triazol-4-ium	Nitrate	437.2 [476]	428.0	−9.2
3-Azido-1-(2-azidoethyl)-1H-1,2,4-triazol-4-ium	5-Nitrotetrazol-1-ide	1138.7 [476]	1194.4	55.7
1-(2-Azidoethyl)-1H-1,2,4-triazol-4-ium	Nitrate	27.2 [476]	92.6	65.4
1-(2-Azidoethyl)-1H-1,2,4-triazol-4-ium	Perchlorate	426.2 [476]	442.3	16.1
3-Azido-1,4-dimethyl-1H-1,2,4-triazol-4-ium	3,5-Dinitro-1,2,4-triazol-1-ide	792.9 [472]	787.3	−5.5
3-Azido-1,4-dimethyl-1H-1,2,4-triazol-4-ium	2,4,6-Trinitrophenolate	611.3 [472]	600.3	−11.0

Table C.2 (continued)

Cation	Anion	Exp.		Equation (2.18)	Dev.
1-(2-Azidoethyl)-1H-1,2,4-triazol-4-ium	5-Nitrotetrazol-1-ide	676.0	[476]	659.6	−16.4
1-(2-Azidoethyl)-1H-1,2,4-triazol-4-ium	4,5-Dinitroimidazol-1-ide	380.2	[476]	376.7	−3.5
1,4-Dimethyl-4H-1,2,4-triazol-1-ium	3,5-Dinitro-1,2,4-triazol-1-ide	118.0	[472]	108.8	−9.2
1,4-Dimethyl-4H-1,2,4-triazol-1-ium	2,4,6-Trinitrophenolate	340.2	[472]	384.6	44.4
4-Amino-1-(2-azidoethyl)-1H-1,2,4-triazol-4-ium	Nitrate	321.1	[476]	327.6	6.5
4-Amino-1-(2-azidoethyl)-1H-1,2,4-triazol-4-ium	Perchlorate	828.4	[476]	828.9	0.5
4-Amino-1-(2-azidoethyl)-1H-1,2,4-triazol-4-ium	3,5-Dinitro-1,2,4-triazol-1-ide	331.7	[477]	333.3	1.6
1-(2-Azidoethyl)-4-methyl-1H-1,2,4-triazol-4-ium	Nitrate	66.3	[476]	38.0	−28.3
1-(2-Azidoethyl)-4-methyl-1H-1,2,4-triazol-4-ium	Perchlorate	568.0	[476]	499.4	−68.6
4-(2-Azidoethyl)-1-methyl-1H-1,2,4-triazol-4-ium	Nitrate	77.7	[476]	38.0	−39.7
4-(2-Azidoethyl)-1-methyl-1H-1,2,4-triazol-4-ium	Perchlorate	541.1	[476]	499.4	−41.7
1-(2-Azidoethyl)-4-methyl-1H-1,2,4-triazol-4-ium	3,5-Dinitro-1,2,4-triazol-1-ide	395.2	[477]	389.5	−5.7
4-(2-Azidoethyl)-1-methyl-1H-1,2,4-triazol-4-ium	3,5-Dinitro-1,2,4-triazol-1-ide	771.3	[477]	732.7	−38.6
4-Methyl-1-propyl-1H-1,2,4-triazol-4-ium	3,5-Dinitro-1,2,4-triazol-1-ide	198.7	[477]	223.0	24.3

Bibliography

[1]   Agrawal JP. High Energy Materials: Propellants, Explosives and Pyrotechnics. John Wiley & Sons; 2010.

[2]   Klapötke TM. Energetic Materials Encyclopedia. Walter de Gruyter GmbH & Co KG; 2018.

[3]   Klapötke TM. Chemistry of High-Energy Materials. Fifth ed. Walter de Gruyter GmbH & Co KG; 2019.

[4]   Keshavarz MH, Klapötke TM. Energetic Compounds: Methods for Prediction of Their Performance. Walter de Gruyter GmbH; 2017.

[5]   Keshavarz MH, Klapötke TM. The Properties of Energetic Materials: Sensitivity, Physical and Thermodynamic Properties. Walter de Gruyter GmbH & Co KG; 2017.

[6]   Keshavarz MH. Combustible Organic Materials: Determination and Prediction of Combustion Properties. Walter de Gruyter GmbH & Co KG; 2018.

[7]   Keshavarz MH. Liquid fuels as jet fuels and propellants. Nova Science; 2018.

[8]   Zhang Q, Shreeve JM. Energetic ionic liquids as explosives and propellant fuels: a new journey of ionic liquid chemistry. Chemical Reviews. 2014;114:10527–74.

[9]   Bastea S, Fried L, Glaesemann K, Howard W, Souers P, Vitello P. CHEETAH 5.0 User's Manual, Lawrence Livermore National Laboratory; 2011.

[10]   Klapötke TM. Chemistry of High-Energy materials. 3th ed. Walter de Gruyter GmbH & Co KG; 2015.

[11]   Keshavarz MH, Abadi YH, Esmaeilpour K, Oftadeh M. Assessment of N-oxide in a new high performance energetic tetrazine derivative on its physical, thermodynamic, sensitivity combustion and detonation performance. Chemistry of Heterocyclic Compounds. 2017;53:797–801.

[12]   Keshavarz MH, Azarniamehraban J, Atabak HH, Ferdowsi M. Recent developments for prediction of power of aromatic and non-aromatic energetic materials along with a novel computer code for prediction of their power. Propellants, Explosives, Pyrotechnics. 2016;41:942–8.

[13]   Roknabadi AG, Keshavarz MH, Esmailpour K, Zamani M. High performance nitroazacubane energetic compounds: Structural, thermochemical and detonation characteristics. ChemistrySelect. 2016;1:6735–40.

[14]   Fotouhi-Far F, Bashiri H, Hamadanian M, Keshavarz MH. A new method for assessment of performing mechanical works of energetic compounds by the cylinder test. Zeitschrift für anorganische und allgemeine Chemie. 2016;642:1086–90.

[15]   Roknabadi AG, Keshavarz MH, Esmailpour K, Zamani M. Structural, thermochemical and detonation performance of derivatives of 1,2,4,5-tetrazine and 1,4 N-oxide 1,2,4,5-tetrazine as new high-performance and nitrogen-rich energetic materials. Journal of the Iranian Chemical Society. 2017;14:57–63.

[16]   Keshavarz MH, Kamalvand M, Jafari M, Zamani A. An improved simple method for the calculation of the detonation performance of CHNOFCl, aluminized and ammonium nitrate explosives. Central European Journal of Energetic Materials. 2016;13:381–96.

[17]   Rezaei AH, Keshavarz MH, Tehrani MK, Darbani SMR, Farhadian AH, Mousavi SJ, Mousaviazar A. Approach for determination of detonation performance and aluminum percentage of aluminized-based explosives by laser-induced breakdown spectroscopy. Applied Optics. 2016;55:3233–40.

[18]   Kim CK, Cho SG, Kim CK, Park HY, Zhang H, Lee HW. Prediction of densities for solid energetic molecules with molecular surface electrostatic potentials. Journal of Computational Chemistry. 2008;29:1818–24.

[19]   Politzer P, Murray JS. Energetic Materials: Part 1. Decomposition, Crystal and Molecular Properties. Elsevier; 2003.

https://doi.org/10.1515/9783110740158-018

[20]   Rice BM, Hare JJ, Byrd EF. Accurate predictions of crystal densities using quantum mechanical molecular volumes. The Journal of Physical Chemistry A. 2007;111:10874–9.

[21]   Qiu L, Xiao H, Gong X, Ju X, Zhu W. Crystal density predictions for nitramines based on quantum chemistry. Journal of Hazardous Materials. 2007;141:280–8.

[22]   Politzer P, Martinez J, Murray JS, Concha MC, Toro-Labbe A. An electrostatic interaction correction for improved crystal density prediction. Molecular Physics. 2009;107:2095–101.

[23]   Tarver CM. Density estimations for explosives and related compounds using the group additivity approach. Journal of Chemical and Engineering Data. 1979;24:136–45.

[24]   Ammon HL. Updated atom/functional group and atom_code volume additivity parameters for the calculation of crystal densities of single molecules, organic salts, and multi-fragment materials containing H, C, B, N, O, F, S, P, Cl, Br, and I. Propellants, Explosives, Pyrotechnics. 2008;33:92–102.

[25]   Willer RL. Calculation of the density and detonation properties of C, H, N, O and F compounds: use in the design and synthesis of new energetic materials. Journal of the Mexican Chemical Society. 2009;53:108–19.

[26]   Keshavarz MH. Prediction of densities of acyclic and cyclic nitramines, nitrate esters and nitroaliphatic compounds for evaluation of their detonation performance. Journal of Hazardous Materials. 2007;143:437–42.

[27]   Keshavarz MH. New method for calculating densities of nitroaromatic explosive compounds. Journal of Hazardous Materials. 2007;145:263–9.

[28]   Keshavarz MH, Pouretedal HR. A reliable simple method to estimate density of nitroaliphatics, nitrate esters and nitramines. Journal of Hazardous Materials. 2009;169:158–69.

[29]   Keshavarz MH. Novel method for predicting densities of polynitro arene and polynitro heteroarene explosives in order to evaluate their detonation performance. Journal of Hazardous Materials. 2009;165:579–88.

[30]   Cho SG, Goh EM, Kim JK. Holographic QSAR models for estimating densities of energetic materials. Bulletin of the Korean Chemical Society. 2001;22:775–8.

[31]   Gorb L, Hill FC, Kholod Y, Muratov EN, Kuz'min VE, Leszczynski J. Progress in Predictions of Environmentally Important Physicochemical Properties of Energetic Materials: Applications of Quantum-Chemical Calculations. In: Practical Aspects of Computational Chemistry II. Springer; 2012. p. 335–59.

[32]   Ahmed A, Sandler SI. Physicochemical properties of hazardous energetic compounds from molecular simulation. Journal of Chemical Theory and Computation. 2013;9:2389–97.

[33]   Katritzky AR, Kuanar M, Slavov S, Hall CD, Karelson M, Kahn I, Dobchev DA. Quantitative correlation of physical and chemical properties with chemical structure: utility for prediction. Chemical Reviews. 2010;110:5714–89.

[34]   Dearden JC, Rotureau P, Fayet G. QSPR prediction of physico-chemical properties for REACH. SAR and QSAR in Environmental Research. 2013;24:279–318.

[35]   Keshavarz MH, Abadi YH, Esmaeilpour K, Damiri S, Oftadeh M. Introducing novel tetrazole derivatives as high performance energetic compounds for confined explosion and as oxidizer in solid propellants. Propellants, Explosives, Pyrotechnics. 2017;42:492–8.

[36]   Keshavarz MH, Abadi YH, Esmaeilpour K, Damiri S, Oftadeh M. A novel class of nitrogen-rich explosives containing high oxygen balance to use as high performance oxidizers in solid propellants. Propellants, Explosives, Pyrotechnics. 2017;42:1155–60.

[37]   Keshavarz MH, Esmaeilpour K, Oftadeh M, Abadi YH. Assessment of two new nitrogen-rich tetrazine derivatives as high performance and safe energetic compounds. RSC Advances. 2015;5:87392–9.

[38]   Keshavarz MH, Abadi YH. Novel organic compounds containing nitramine groups suitable as high-energy cyclic nitramine compounds. ChemistrySelect. 2018;3:8238–44.

[39] Keshavarz MH, Abadi YH, Esmaeilpour K, Damiri S, Oftadeh M. Assessment of the effect of N-oxide group in a new high-performance energetic tetrazine derivative on its physicochemical and thermodynamic properties, sensitivity, and combustion and detonation performance. Chemistry of Heterocyclic Compounds. 2017;53:797–801.

[40] Keshavarz MH, Abadi YH, Esmaeilpour K, Damiri S, Oftadeh M. Novel high-nitrogen content energetic compounds with high detonation and combustion performance for use in plastic bonded explosives (PBXs) and composite solid propellants. Central European Journal of Energetic Materials. 2018;15:364–75.

[41] Keshavarz MH, Esmailpour K, Zamani M, Roknabadi AG. Thermochemical, sensitivity and detonation characteristics of new thermally stable high performance explosives. Propellants, Explosives, Pyrotechnics. 2015;40:886–91.

[42] Wang Q, Shao Y, Lu M. Azo1,3,4-oxadiazole as a Novel Building Block to Design High-Performance Energetic Materials. Crystal Growth & Design. 2019;19:839–44.

[43] Fershtat LL, Ovchinnikov IV, Epishina MA, Romanova AA, Lempert DB, Muravyev NV, Makhova NN. Assembly of nitrofurazan and nitrofuroxan frameworks for high-performance energetic materials. ChemPlusChem. 2017;82:1315–9.

[44] Dalinger IL, Kormanov AV, Suponitsky KY, Muravyev NV, Sheremetev AB. Pyrazole–tetrazole hybrid with trinitromethyl, fluorodinitromethyl, or (difluoroamino) dinitromethyl groups: high-performance energetic materials. Chemistry–An Asian Journal. 2018;13:1165–72.

[45] Sun S, Lu M. Conjugation in multi-tetrazole derivatives: a new design direction for energetic materials. Journal of Molecular Modeling. 2018;24:173.

[46] Yang J, Gong X, Mei H, Li T, Zhang J, Gozin M. Design of zero oxygen balance energetic materials on the basis of Diels–Alder chemistry. The Journal of Organic Chemistry. 2018;83:14698–702.

[47] Ammon HL. New atom/functional group volume additivity data bases for the calculation of the crystal densities of C-, H-, N-, O-, F-, S-, P-, Cl-, and Br-containing compounds. Structural Chemistry. 2001;12:205–12.

[48] Ye C, Shreeve JM. New atom/group volume additivity method to compensate for the impact of strong hydrogen bonding on densities of energetic materials. Journal of Chemical & Engineering Data. 2008;53:520–4.

[49] Smirnov AS, Smirnov SP, Pivina TS, Lempert DB, Maslova LK. Comprehensive assessment of physicochemical properties of new energetic materials. Russian Chemical Bulletin. 2016;65:2315–32.

[50] Hofmann DWM. Fast estimation of crystal densities. Acta Crystallographica Section B: Structural Science. 2002;58:489–93.

[51] Ghule VD, Nirwan A, Devi A. Estimating the densities of benzene-derived explosives using atomic volumes. Journal of Molecular Modeling. 2018;24:50.

[52] Ye C, Shreeve JM. Rapid and accurate estimation of densities of room-temperature ionic liquids and salts. The Journal of Physical Chemistry A. 2007;111:1456–61.

[53] Beaucamp S, Marchet N, Mathieu D, Agafonov V. Calculation of the crystal densities of molecular salts and hydrates using additive volumes for charged groups. Acta Crystallographica Section B: Structural Science. 2003;59:498–504.

[54] Meyer R, Köhler J, Homburg A. Explosives. 6th ed. John Wiley & Sons; 2008.

[55] Hong D, Li Y, Zhu S, Zhang L, Pang C. Three insensitive energetic co-crystals of 1-nitronaphthalene, with 2,4,6-trinitrotoluene (TNT), 2,4,6-trinitrophenol (picric acid) and D-mannitol hexanitrate (MHN). Central European Journal of Energetic Materials. 2015;12:47–62.

[56] Rice BM, Sorescu DC. Assessing a generalized CHNO intermolecular potential through ab initio crystal structure prediction. The Journal of Physical Chemistry B. 2004;108:17730–9.

[57]   Murray JS, Brinck T, Politzer P. Relationships of molecular surface electrostatic potentials to some macroscopic properties. Chemical Physics. 1996;204:289–99.

[58]   Pan J-F, Lee Y-W. Crystal density prediction for cyclic and cage compounds. Physical Chemistry Chemical Physics. 2004;6:471–3.

[59]   Kim CK, Lee KA, Hyun KH, Park HJ, Kwack IY, Kim CK, Lee HW, Lee BS. Prediction of physicochemical properties of organic molecules using van der Waals surface electrostatic potentials. Journal of Computational Chemistry. 2004;25:2073–9.

[60]   Moxnes JF, Hansen FK, Jensen TL, Sele ML, Unneberg E. A computational study of density of some high energy molecules. Propellants, Explosives, Pyrotechnics. 2017;42:204–12.

[61]   Kim CK, Cho SG, Kim CK, Kim M-R, Lee HW. Prediction of physicochemical properties of organic molecules using semi-empirical methods. Bulletin of the Korean Chemical Society. 2013;34:1043.

[62]   Rice BM, Byrd EFC. Evaluation of electrostatic descriptors for predicting crystalline density. Journal of Computational Chemistry. 2013;34:2146–51.

[63]   Kim CK, Cho SG, Li J, Park BH, Kim CK. Prediction of crystal density and explosive performance of high-energy-density molecules using the modified MSEP ccheme. Bulletin of the Korean Chemical Society. 2016;37:1683–9.

[64]   Wang G, Xu Y, Xue C, Ding Z, Liu Y, Liu H, Gong X. Prediction of the Crystalline Densities of Aliphatic Nitrates by Quantum Chemistry Methods. Central European Journal of Energetic Materials. 2019;16:412–32.

[65]   Wang L, Zhang M, Chen J, Su L, Zhao S, Zhang C, Liu J, Chen C. Corrections of Molecular Morphology and Hydrogen Bond for Improved Crystal Density Prediction. Molecules. 2020;25:161.

[66]   Keshavarz MH, Jafari M, Ebadpour R. Recent advances for assessment of the condensed phase heat of formation of high-energy content organic compounds and ionic liquids (or salts) to introduce a new computer code for design of desirable compounds. Fluid Phase Equilibria. 2020;533:112913.

[67]   Nirwan A, Devi A, Ghule VD. Assessment of density prediction methods based on molecular surface electrostatic potential. Journal of Molecular Modeling. 2018;24:166.

[68]   Ghule VD, Nirwan A. Role of forcefield in density prediction for CHNO explosives. Structural Chemistry. 2018;29:1375–82.

[69]   Keshavarz MH, Klapötke TM, Sućeska M. Energetic materials designing bench (EMDB), version 1.0. Propellants, Explosives, Pyrotechnics. 2017;42:854–6.

[70]   Fathollahi M, Sajady H. Prediction of density of energetic cocrystals based on QSPR modeling using artificial neural network. Structural Chemistry. 2018;29:1119–28.

[71]   Rahimi R, Keshavarz MH, Akbarzadeh AR. Prediction of the density of energetic materials on the basis of their molecular structures. Central European Journal of Energetic Materials. 2016;13:73–101.

[72]   Pagoria PF, Lee GS, Mitchell AR, Schmidt RD. A review of energetic materials synthesis. Thermochimica Acta. 2002;384:187–204.

[73]   Keshavarz MH, Soury H, Motamedoshariati H, Dashtizadeh A. Improved method for prediction of density of energetic compounds using their molecular structure. Structural Chemistry. 2015;26:455–66.

[74]   Zohari N, Sheibani N. Link between density and molecular structures of energetic azido compounds as green plasticizers. Zeitschrift für anorganische und allgemeine Chemie. 2016;642:1472–9.

[75]   Yang J, Zhang X, Gao P, Gong X, Wang G. Exploring highly energetic aliphatic azido nitramines for plasticizers. RSC Advances. 2014;4:53172–9.

[76]   Keshavarz MH, Rahimi R, Akbarzadeh AR. Two novel correlations for assessment of crystal density of hazardous ionic molecular energetic materials using their molecular structures. Fluid Phase Equilibria. 2015;402:1–8.

[77]   Zohari N, Abrishami F, Ebrahimikia M. Investigation of the Effect of Various Substituents on the Density of Tetrazolium Nitrate Salts as Green Energetic Materials. Zeitschrift für anorganische und allgemeine Chemie. 2016;642:749–60.

[78]   Zohari N, Bajestani IR. A novel correlation for predicting the density of tetrazole–N-oxide salts as green Energetic Materials through Their molecular structure. Central European Journal of Energetic Materials. 2018;15:629–51.

[79]   He P, Zhang JG, Yin X, Wu JT, Wu L, Zhou ZN, Zhang TL. Chemistry—A European Journal, Chem Eur J. 2016;22:7670–85.

[80]   Agrawal JP, Hodgson R. Organic Chemistry of Explosives. John Wiley & Sons; 2007.

[81]   Sikder AK, Maddala G, Agrawal JP, Singh H. Important aspects of behaviour of organic energetic compounds: a review. Journal of Hazardous Materials. 2001;84:1–26.

[82]   Sikder A, Sikder N. A review of advanced high performance, insensitive and thermally stable energetic materials emerging for military and space applications. Journal of Hazardous Materials. 2004;112:1–15.

[83]   Keshavarz MH. Research progress on heats of formation and detonation of energetic compounds. In: Brar SK, editor. Hazardous Materials: Types, Risks and Control. New York, New York: Nova Science Publishers Inc.; 2011. p. 339–59.

[84]   Rogers DW, Zavitsas AA, Matsunaga N. Determination of enthalpies ('Heats') of formation. Wiley Interdisciplinary Reviews: Computational Molecular Science. 2013;3:21–36.

[85]   Goodwin A, Marsh K, Wakeham W. Measurement of the thermodynamic properties of single phases. Elsevier; 2003.

[86]   Sućeska M. EXPLO5–Computer program for calculation of detonation parameters. In: Proc. of 32nd Int. Annual Conference of ICT. Karlsruhe, Germany. 2001.

[87]   Keshavarz MH, Motamedoshariati H, Moghayadnia R, Nazari HR, Azarniamehraban J. A new computer code to evaluate detonation performance of high explosives and their thermochemical properties, part I. Journal of Hazardous Materials. 2009;172:1218–28.

[88]   Keshavarz MH, Motamedoshariati H, Moghayadnia R, Ghanbarzadeh M, Azarniamehraban J. A new computer code for assessment of energetic materials with crystal density, condensed phase enthalpy of formation, and activation energy of thermolysis. Propellants, Explosives, Pyrotechnics. 2013;38:95–102.

[89]   Keshavarz MH, Ghani K, Asgari A. A suitable computer code for prediction of sublimation energy and deflagration temperature of energetic materials. Journal of Thermal Analysis and Calorimetry. 2015;121:675–81.

[90]   Gordon S, McBride BJ. Computer program for calculation of complex chemical equilibrium compositions and applications. Part 1: Analysis. Washington DC: NASA Lewis Research Center; 1994.

[91]   Keshavarz MH, Tehrani MK, Pouretedal HR, Semnani A. New pathway for quick estimation of gas phase heat of formation of non-aromatic energetic compounds. Indian Journal of Engineering and Materials Sciences. 2006;13:542–8.

[92]   Keshavarz MH, Tehrani MK. A new method for determining gas phase heat of formation of aromatic energetic compounds. Propellants, Explosives, Pyrotechnics. 2007;32:155–9.

[93]   Gharagheizi F, Sattari M, Tirandazi B. Prediction of crystal lattice energy using enthalpy of sublimation: a group contribution-based model. Industrial & Engineering Chemistry Research. 2011;50:2482–6.

[94]   Mathieu D. Simple alternative to neural networks for predicting sublimation enthalpies from fragment contributions. Industrial & Engineering Chemistry Research. 2012;51:2814–9.

[95]   Meftahi N, Walker ML, Enciso M, Smith BJ. Predicting the enthalpy and Gibbs energy of sublimation by QSPR modeling. Scientific Reports. 2018;8:1–9.

[96]   Suntsova MA, Dorofeeva OV. Prediction of enthalpies of sublimation of high-nitrogen energetic compounds: Modified Politzer model. Journal of Molecular Graphics and Modelling. 2017;72:220–8.

[97] Keshavarz MH, Yousefi MH. Heats of sublimation of nitramines based on simple parameters. Journal of Hazardous Materials. 2008;152:929–33.

[98] Keshavarz MH. Prediction of heats of sublimation of nitroaromatic compounds via their molecular structure. Journal of Hazardous Materials. 2008;151:499–506.

[99] Keshavarz MH. Improved prediction of heats of sublimation of energetic compounds using their molecular structure. Journal of Hazardous Materials. 2010;177:648–59.

[100] Keshavarz MH, Bashavard B, Goshadro A, Dehghan Z, Jafari M. Prediction of heats of sublimation of energetic compounds using their molecular structures. Journal of Thermal Analysis and Calorimetry. 2015;120:1941–51.

[101] Sana M, Leroy G, Peeters D, Wilante C. The theoretical study of the heats of formation of organic compounds containing the substituents CH3, CF3, NH2, NF2, NO2, OH and F. Journal of Molecular Structure: THEOCHEM. 1988;164:249–74.

[102] Sprague JT, Tai JC, Yuh Y, Allinger NL. The MMP2 calculational method. Journal of Computational Chemistry. 1987;8:581–603.

[103] Lii JH, Allinger NL. Molecular mechanics. The MM3 force field for hydrocarbons. 3. The van der Waals' potentials and crystal data for aliphatic and aromatic hydrocarbons. Journal of the American Chemical Society. 1989;111:8576–82.

[104] Nevins N, Lii JH, Allinger NL. Molecular mechanics (MM4) calculations on conjugated hydrocarbons. Journal of Computational Chemistry. 1996;17:695–729.

[105] Allinger NL. Molecular Structure: Understanding Steric and Electronic Effects from Molecular Mechanics. John Wiley & Sons; 2010.

[106] Akutsu Y, Tahara S-Y, Tamura M, Yoshida T. Calculations of heats of formation for nitro compounds by semi-empirical MO methods and molecular mechanics. Journal of Energetic Materials. 1991;9:161–71.

[107] Mole SJ, Zhou X, Liu R. Density functional theory (DFT) study of enthalpy of formation. 1. Consistency of DFT energies and atom equivalents for converting DFT energies into enthalpies of formation. The Journal of Physical Chemistry. 1996;100:14665–71.

[108] Dewar MJ, Zoebisch EG, Healy EF, Stewart JJ. Development and use of quantum mechanical molecular models. 76. AM1: a new general purpose quantum mechanical molecular model. Journal of the American Chemical Society. 1985;107:3902–9.

[109] Stewart JJP. Optimization of parameters for semiempirical methods II. Applications, Journal of Computational Chemistry. 1989;10:221–64.

[110] Stewart JJP. Optimization of parameters for semiempirical methods V: modification of NDDO approximations and application to 70 elements. Journal of Molecular Modeling. 2007;13:1173–213.

[111] Stewart JJP. Optimization of parameters for semiempirical methods VI: more modifications to the NDDO approximations and re-optimization of parameters. Journal of Molecular Modeling. 2013;19:1–32.

[112] Chen C, Wu J. Correlations between theoretical and experimental determination of heat of formation of certain aromatic nitro compounds. Computers & Chemistry. 2001;25:117–24.

[113] Oftadeh M, Keshavarz MH, Khodadadi R. Prediction of the condensed phase enthalpy of formation of nitroaromatic compounds using the estimated gas phase enthalpies of formation by the PM3 and B3LYP methods. Central European Journal of Energetic Materials. 2014;11:143–56.

[114] Chickos JS, Gavezzotti A. Sublimation enthalpies of organic compounds: a very large database with a match to crystal structure determinations and a comparison with lattice energies. Crystal Growth & Design. 2019;19:6566–76.

[115] Atkins PW, De Paula J, Keeler J. Atkins' Physical Chemistry. Eleventh ed. Oxford university press; 2018.

[116] Byrd EFC, Rice BM. Improved prediction of heats of formation of energetic materials using quantum mechanical calculations. The Journal of Physical Chemistry A. 2006;110:1005–13.

[117] Ohlinger WS, Klunzinger PE, Deppmeier BJ, Hehre WJ. Efficient calculation of heats of formation. The Journal of Physical Chemistry A. 2009;113:2165–75.

[118] Politzer P, Murray JS. Quantitative treatments of solute/solvent interactions. Elsevier Science; 1994.

[119] Rice BM, Pai SV, Hare J. Predicting heats of formation of energetic materials using quantum mechanical calculations. Combustion and Flame. 1999;118:445–58.

[120] Rice BM, Hare J. Predicting heats of detonation using quantum mechanical calculations. Thermochimica Acta. 2002;384:377–91.

[121] Politzer P, Murray JS, Edward Grice M, Desalvo M, Miller E. Calculation of heats of sublimation and solid phase heats of formation. Molecular Physics. 1997;91:923–8.

[122] Pan Y, Zhu W, Xiao H. Molecular design on a new family of azaoxaadamantane cage compounds as potential high-energy density compounds. Canadian Journal of Chemistry. 2019;97:86–93.

[123] Khan RU, Zhu W. Designing and looking for novel low-sensitivity and high-energy cage derivatives based on the skeleton of nonanitro nonaaza pentadecane framework. Structural Chemistry. 2020;31:1387–402.

[124] Du M, Han T, Liu F, Wu H. Theoretical investigation of the structure, detonation properties, and stability of bicyclo [3.2.1] octane derivatives. Journal of Molecular Modeling. 2019;25:253.

[125] Duan B, Liu N, Wang B, Lu X, Mo H. Comparative theoretical studies on a series of novel energetic salts composed of 4,8-dihydrodifurazano [3,4-b,e] pyrazine-based anions and ammonium-based cations. Molecules. 2019;24:3213.

[126] Li X-H, Zhang C, Ju X-H. Theoretical screening of bistriazole-derived energetic salts with high energetic properties and low sensitivity. RSC advances. 2019;9:26442–9.

[127] Zhao S-X, Xia Q-Y, Zhang C, Xing X-L, Ju X-H. Theoretical design of bistetrazole diolate derivatives as novel non-nitro energetic salts with low sensitivity. Structural Chemistry. 2019;30:1015–22.

[128] de Oliveira RSS, Borges I. Correlation between molecular charge densities and sensitivity of nitrogen-rich heterocyclic nitroazole derivative explosives. Journal of Molecular Modeling. 2019;25:314.

[129] Jaidann M, Roy S, Abou-Rachid H, Lussier L-S. A DFT theoretical study of heats of formation and detonation properties of nitrogen-rich explosives. Journal of Hazardous Materials. 2010;176:165–73.

[130] Poling BE, Prausnitz JM, Paul OCJ, Reid RC. The Properties of Gases and Liquids. 5th ed. New York: McGraw-Hill; 2001.

[131] Argoub K, Benkouider AM, Yahiaoui A, Kessas R, Guella S, Bagui F. Prediction of standard enthalpy of formation in the solid state by a third-order group contribution method. Fluid Phase Equilibria. 2014;380:121–7.

[132] Salmon A, Dalmazzone D. Prediction of enthalpy of formation in the solid state (at 298.15 K) using second-order group contributions—Part 2: Carbon-cydrogen, carbon-hydrogen-oxygen, and carbon-hydrogen-nitrogen-oxygen compounds. Journal of Physical and Chemical Reference Data. 2007;36:19–58.

[133] Bourasseau S. A systematic procedure for estimating the standard heats of formation in the condensed state of non aromatic polynitro-compounds. Journal of Energetic Materials. 1990;8:266–91.

[134] Jafari M, Keshavarz MH, Noorbala MR, Kamalvand M. A Reliable Method for Prediction of the Condensed Phase Enthalpy of Formation of High Nitrogen Content Materials through their Gas Phase Information. ChemistrySelect. 2016;1:5286–96.

[135] Keshavarz MH, Oftadeh M. New method for estimating the heat of formation of CHNO explosives in crystalline state. High Temperatures-High Pressures. 2004;36:499–504.

[136] Keshavarz MH, Sadeghi H. A new approach to predict the condensed phase heat of formation in acyclic and cyclic nitramines, nitrate esters and nitroaliphatic energetic compounds. Journal of Hazardous Materials. 2009;171:140–6.

[137] Keshavarz MH. Predicting condensed phase heat of formation of nitroaromatic compounds. Journal of Hazardous Materials. 2009;169:890–900.

[138] Keshavarz MH. Prediction of the condensed phase heat of formation of energetic compounds. Journal of Hazardous Materials. 2011;190:330–44.

[139] Keshavarz MH, Oftadeh M. New method for estimating the heat of formation of CHNO explosives in crystalline state. High Temperatures-High Pressures. 2004;35:499–504.

[140] Kamlet MJ, Jacobs SJ. The chemistry of detonation. 1. A simple method for calculating detonation properties of CHNO explosives. Journal of Chemical Physics. 1967;48:23–35.

[141] Keshavarz MH. A simple procedure for calculating condensed phase heat of formation of nitroaromatic energetic materials. Journal of Hazardous Materials. 2006;136:425–31.

[142] Afeefy H, Liebman J, Stein S, Linstrom P, Mallard W. NIST Chemistry WebBook, NIST Standard Reference Database Number 69, Linstrom PJ, Mallard WG, editors; 2011.

[143] Keshavarz MH. Theoretical prediction of condensed phase heat of formation of nitramines, nitrate esters, nitroaliphatics and related energetic compounds. Journal of Hazardous Materials. 2006;136:145–50.

[144] Nazari B, Keshavarz MH, Hamadanian M, Mosavi S, Ghaedsharafi AR, Pouretedal HR. Reliable prediction of the condensed (solid or liquid) phase enthalpy of formation of organic energetic materials at 298 K through their molecular structures. Fluid Phase Equilibria. 2016;408:248–58.

[145] Rice BM, Byrd EFC, Mattson WD. Computational aspects of nitrogen-rich HEDMs. In: High Energy Density Materials. Springer; 2007. p. 153–94.

[146] Wei H, Zhang J, Shreeve JnM. Synthesis, characterization, and energetic properties of 6-amino-tetrazolo [1,5-b]-1,2,4,5-tetrazine-7-N-oxide: A nitrogen-rich material with high density. Chemistry–An Asian Journal. 2015;10:1130–2.

[147] Chavez DE, Hiskey MA. 1,2,4,5-Tetrazine based energetic materials. Journal of Energetic Materials. 1999;17:357–77.

[148] Klapötke TM. The Synthesis Chemistry of Energetic Materials. In: Armstrong RW, editor. Energetics Science and Technology in Central Europe. Maryland: CALCE EPSC Press; 2012. p. 57.

[149] Talawar MB, Sivabalan R, Asthana SN, Singh H. Novel ultrahigh-energy materials. Combustion, Explosion and Shock waves. 2005;41:264–77.

[150] Damse R, Ghosh M, Naik N, Sikder A. Thermoanalytical screening of nitrogen-rich compounds for ballistic requirements of gun propellant. Journal of Propulsion and Power. 2009;25:249–56.

[151] Nair U, Asthana S, Rao AS, Gandhe B. Advances in High Energy Materials (Review Paper). Defence Science Journal. 2010;60:137–51.

[152] Gao H, Shreeve JnM. Azole-based energetic salts. Chemical Reviews. 2011;111:7377–436.

[153] Jafari M, Keshavarz MH. Simple approach for predicting the heats of formation of high nitrogen content materials. Fluid Phase Equilibria. 2016;415:166–75.

[154] Huynh MHV, Hiskey MA, Chavez DE, Naud DL, Gilardi RD. Synthesis, characterization, and energetic properties of diazido heteroaromatic high-nitrogen C–N compound. Journal of the American Chemical Society. 2005;127:12537–43.

[155] Williams MM, McEwan WS, Henry RA. The heats of combustion of substituted triazoles, tetrazoles and related high nitrogen compounds. The Journal of Physical Chemistry. 1957;61:261–7.

[156] Zhang Q, Shreeve JnM. Energetic ionic liquids as explosives and propellant fuels: a new journey of ionic liquid chemistry. Chemical Reviews. 2014;114:10527–74.

[157] Sebastiao E, Cook C, Hu A, Murugesu M. Recent developments in the field of energetic ionic liquids. Journal of Materials Chemistry A. 2014;2:8153–73.

[158] Thomas E, Vijayalakshmi KP, George BK. Imidazolium based energetic ionic liquids for monopropellant applications: a theoretical study. RSC Advances. 2015;5:71896–902.

[159] Singh HJ, Mukherjee U. A computational approach to design energetic ionic liquids. Journal of Molecular Modeling. 2013;19:2317–27.

[160] Bhosale VK, Kulkarni PS. Hypergolic behavior of pyridinium salts containing cyanoborohydride and dicyanamide anions with oxidizer RFNA. Propellants, Explosives, Pyrotechnics. 2016;41:1013–9.

[161] Nimesh S, Ang HG. 1-(2H-tetrazolyl)-1,2,4-triazole-5-amine (TzTA)—a thermally stable nitrogen rich energetic material: synthesis, characterization and thermo-chemical analysis. Propellants, Explosives, Pyrotechnics. 2015;40:426–32.

[162] Keshavarz MH, Klapötke TM. Energetic Compounds: Methods for Prediction of Their Performance. 2nd ed. Walter de Gruyter GmbH & Co KG; 2020.

[163] Nirwan A, Ghule VD. Estimation of heats of formation for nitrogen-rich cations using G3, G4, and G4 (MP2) theoretical methods. Theoretical Chemistry Accounts. 2018;137:115.

[164] Gao H, Ye C, Piekarski CM, Shreeve JM. Computational characterization of energetic salts. The Journal of Physical Chemistry C. 2007;111:10718–31.

[165] Jenkins HDB, Tudela D, Glasser L. Lattice potential energy estimation for complex ionic salts from density measurements. Inorganic Chemistry. 2002;41:2364–7.

[166] Zhang X, Zhu W, Wei T, Zhang C, Xiao H. Densities, heats of formation, energetic properties, and thermodynamics of formation of energetic nitrogen-rich salts containing substituted protonated and methylated tetrazole cations: A computational study. The Journal of Physical Chemistry C. 2010;114:13142–52.

[167] Keshavarz MH, Nazari B, Jafari M, Bakhtiari R. A simple approach for prediction of the condensed phase heat of formation of imidazolium-based ionic liquids or salts. ChemistrySelect. 2018;3:3505–10.

[168] Jafari M, Davtalab M, Keshavarz MH, Esmaeilpour K, Mosaviazar A, Ghasemi MA, Amini M. Accurate prediction of the condensed phase (solid or liquid) heat of formation of triazolium-based energetic ionic salts at 298.15 K. Central European Journal of Energetic Materials. 2018;15:501–15.

[169] Dong L-L, He L, Liu H-Y, Tao G-H, Nie F-D, Huang M, Hu C-W. Nitrogen-Rich Energetic Ionic Liquids Based on the N,N-Bis(1H-tetrazol-5-yl)amine Anion – Syntheses, Structures, and Properties. European Journal of Inorganic Chemistry. 2013;2013:5009–19.

[170] Xue H, Twamley B, Jeanne MS. Energetic salts of substituted 1,2,4-triazolium and tetrazolium 3,5-dinitro-1,2,4-triazolates. Journal of Materials Chemistry. 2005;15:3459–65.

[171] Berthod A, Ruiz-Ángel M, Carda-Broch S. Recent advances on ionic liquid uses in separation techniques. Journal of Chromatography A. 2018;1559:2–16.

[172] Al-Fakih A, Algamal Z, Lee M, Aziz M. A penalized quantitative structure–property relationship study on melting point of energetic carbocyclic nitroaromatic compounds using adaptive bridge penalty. SAR and QSAR in Environmental Research. 2018;29:339–53.

[173] Yagofarov MI, Solomonov BN. Calculation of the fusion enthalpy temperature dependence of polyaromatic hydrocarbons from the molecular structure: old and new approaches. The Journal of Chemical Thermodynamics. 2021;152:106278.

[174] Liu Y, Lai W, Yu T, Ma Y, Guo W, Ge Z. Melting point prediction of energetic materials via continuous heating simulation on solid-to-liquid phase transition. ACS Omega. 2019;4:4320–4.

[175] Joback KG, Reid RC. Estimation of pure-component properties from group-contributions. Chemical Engineering Communications. 1987;57:233–43.

[176] Keshavarz MH, Pouretedal HR. New approach for predicting melting point of carbocyclic nitroaromatic compounds. Journal of Hazardous Materials. 2007;148:592–8.

[177] Lydersen A, Greenkorn RA, Hougen OA. Estimation of Critical Properties of Organic Compounds by the Method of Group Contibutions. University of Wisconsin; 1955.

[178] Ambrose D. Correlation and Estimation of Vapour-Liquid Critical Properties. National Physical Library; 1978.

[179] Klincewicz KM, Reid RC. Estimation of critical properties with group contribution methods. AIChE Journal. 1984;30:137–42.

[180] Lyman WJ, Reehl WF, Rosenblatt DH, Rosenblatt DH. Handbook of Chemical Property Estimation Methods: Environmental Behavior of Organic Compounds. New York: McGraw-Hill; 1982.

[181] Constantinou L, Prickett SE, Mavrovouniotis ML. Estimation of thermodynamic and physical properties of acyclic hydrocarbons using the ABC approach and conjugation operators. Industrial & Engineering Chemistry Research. 1993;32:1734–46.

[182] Prickett SE, Constantinou L, Mavrovouniotis ML. Computational identification of conjugate paths for estimation of properties of organic compounds. Molecular Simulation. 1993;11:205–28.

[183] Constantinou L, Gani R. New group contribution method for estimating properties of pure compounds. AIChE Journal. 1994;40:1697–710.

[184] Constantinou L, Prickett SE, Mavrovouniotis ML. Estimation of properties of acyclic organic compounds using conjugation operators. Industrial & Engineering Chemistry Research. 1994;33:395–402.

[185] Marrero-Morejón J, Pardillo-Fontdevila E. Estimation of pure compound properties using group-interaction contributions. AIChE journal. 1999;45:615–21.

[186] Marrero J, Gani R. Group-contribution based estimation of pure component properties. Fluid Phase Equilibria. 2001;183:183–208.

[187] Simamora P, Yalkowsky SH. Group contribution methods for predicting the melting points and boiling points of aromatic compounds. Industrial & Engineering Chemistry Research. 1994;33:1405–9.

[188] Krzyzaniak JF, Myrdal PB, Simamora P, Yalkowsky SH. Boiling point and melting point prediction for aliphatic, non-hydrogen-bonding compounds. Industrial & Engineering Chemistry Research. 1995;34:2530–5.

[189] Alamdari RF, Keshavarz MH. A simple method to predict melting points of non-aromatic energetic compounds. Fluid Phase Equilibria. 2010;292:1–6.

[190] Agrawal PM, Rice BM, Thompson DL. Molecular dynamics study of the melting of nitromethane. The Journal of Chemical Physics. 2003;119:9617–27.

[191] Alavi S, Thompson DL. Simulations of the solid, liquid, and melting of 1-n-butyl-4-amino-1,2,4-triazolium bromide. The Journal of Physical Chemistry B. 2005;109:18127–34.

[192] Agrawal PM, Rice BM, Zheng L, Velardez GF, Thompson DL. Molecular dynamics simulations of the melting of 1,3,3-trinitroazetidine. Journal of Physical Chemistry B. 2006;110:5721.

[193] Siavosh-Haghighi A, Thompson DL. Melting point determination from solid–liquid coexistence initiated by surface melting. The Journal of Physical Chemistry C. 2007;111:7980–5.

[194] Jain A, Yalkowsky SH. Estimation of melting points of organic compounds-II. Journal of Pharmaceutical Sciences. 2006;95:2562–618.

[195] Jain A, Yalkowsky SH. Comparison of two methods for estimation of melting points of organic compounds. Industrial & Engineering Chemistry Research. 2007;46:2589–92.

[196] Evans DC, Yalkowsky SH. A simplified prediction of entropy of melting for energetic compounds. Fluid Phase Equilibria. 2011;303:10–4.

[197] Hilal SH, Carreira LA, Karickhoff SW. Estimation of chemical reactivity parameters and physical properties of organic molecules using SPARC. In: Murry PPaJS, editor. Theoretical and Computational Chemistry. Amsterdam: Elsevier; 1994. p. 291–353.

[198] Whiteside T, Hilal S, Brenner A, Carreira L. Estimating the melting point, entropy of fusion, and enthalpy of fusion of organic compounds via SPARC. SAR and QSAR in Environmental Research. 2016;27:677–701.

[199] Admire B, Lian B, Yalkowsky SH. Estimating the physicochemical properties of polyhalogenated aromatic and aliphatic compounds using UPPER: Part 1. Boiling point and melting point. Chemosphere. 2015;119:1436–40.

[200] Lian B, Yalkowsky SH. Unified physicochemical property estimation relationships (UPPER). Journal of Pharmaceutical Sciences. 2014;103:2710–23.

[201] Admire B, Lian B, Yalkowsky SH. Estimating the physicochemical properties of polyhalogenated aromatic and aliphatic compounds using UPPER: Part 2. Aqueous solubility, octanol solubility and octanol–water partition coefficient. Chemosphere. 2015;119:1441–6.

[202] Alantary D, Yalkowsky SH. Estimating the physicochemical properties of polysubstituted aromatic compounds using UPPER. Journal of Pharmaceutical Sciences. 2018;107:297–306.

[203] Brown RJC, Brown RFC. Melting point and molecular symmetry. Journal of Chemical Education. 2000;77:724.

[204] Carnelley T. XIII. Chemical symmetry, or the influence of atomic arrangement on the physical properties of compounds. The London, Edinburgh, and Dublin Philosophical Magazine and Journal of Science. 1882;13:112–30.

[205] Keshavarz MH. New method for predicting melting points of polynitro arene and polynitro heteroarene compounds. Journal of Hazardous Materials. 2009;171:786–96.

[206] Agrawal JP. Recent trends in high-energy materials. Progress in Energy and Combustion Science. 1998;24:1–30.

[207] Keshavarz MH. Approximate prediction of melting point of nitramines, nitrate esters, nitrate salts and nitroaliphatics energetic compounds. Journal of Hazardous Materials. 2006;138:448–51.

[208] Keshavarz MH, Gharagheizi F, Pouretedal HR. Improved reliable approach to predict melting points of energetic compounds. Fluid Phase Equilibria. 2011;308:114–28.

[209] Khozani MH, Keshavarz MH, Nazari B, Mohebbi M. Simple approach for prediction of melting points of organic molecules containing hazardous peroxide bonds. Journal of the Iranian Chemical Society. 2015;12:587–98.

[210] Nazari B, Hamadanian M, Keshavarz MH, Rezaei J. New method for assessment of melting points of organic azides using their molecular structures. Fluid Phase Equilibria. 2016;427:27–34.

[211] Hodgson HH, Norris W. Replacement of the diazonium by the azido group in acid solution. Journal of the Chemical Society (Resumed). 1949;162:762–3.

[212] Hamadanian M, Keshavarz MH, Nazari B, Mohebbi M. Reliable method for safety assessment of melting points of energetic compounds. Process Safety and Environmental Protection. 2016;103:10–22.

[213] Keshavarz MH, Maghsoodi NK, Shokrollahi A. A reliable model for assessment of melting points of cyclic hydrocarbons containing complex molecular structures, isomers and stereoisomers. Fluid Phase Equilibria. 2020;521:112692.

[214] Singh SK, Savoy AW. Ionic liquids synthesis and applications: An overview. Journal of Molecular Liquids. 2020;297:112038.

[215] Bera A, Agarwal J, Shah M, Shah S, Vij RK. Recent advances in ionic liquids as alternative to surfactants/chemicals for application in upstream oil industry. Journal of Industrial and Engineering Chemistry. 2020;82:17–30.

[216] Nasirpour N, Mohammadpourfard M, Heris SZ. Ionic liquids: Promising compounds for sustainable chemical processes and applications. Chemical Engineering Research and Design. 2020;160:264–300.

[217] Qin M, Zhong F, Sun Y, Tan X, Hu K, Zhang H, Kong M, Wang G, Zhuang L. Effect of cation substituent of dodecanesulfate-based anionic surface active ionic liquids on micellization: Experimental and theoretical studies. Journal of Molecular Liquids. 2020;303:112695.

[218] Low K, Kobayashi R, Izgorodina EI. The effect of descriptor choice in machine learning models for ionic liquid melting point prediction. The Journal of Chemical Physics. 2020;153:104101.

[219] Venkatraman V, Evjen S, Knuutila HK, Fiksdahl A, Alsberg BK. Predicting ionic liquid melting points using machine learning. Journal of Molecular Liquids. 2018;264:318–26.

[220] Chen L, Bryantsev VS. A density functional theory based approach for predicting melting points of ionic liquids. Physical Chemistry Chemical Physics. 2017;19:4114–24.

[221] Mehrkesh A, Karunanithi AT. New quantum chemistry-based descriptors for better prediction of melting point and viscosity of ionic liquids. Fluid Phase Equilibria. 2016;427:498–503.

[222] Lazzús JA. A group contribution method for predicting the freezing point of ionic liquids. Periodica Polytechnica Chemical Engineering. 2016;60:273–81.

[223] Zhang S, Lu X, Zhou Q, Li X, Zhang X, Li S. Ionic Liquids: Physicochemical Properties. Elsevier; 2009.

[224] Todeschini R, Consonni V. Handbook of Molecular Descriptors. John Wiley & Sons; 2008.

[225] Wang D, Yuan Y, Duan S, Liu R, Gu S, Zhao S, Liu L, Xu J. QSPR study on melting point of carbocyclic nitroaromatic compounds by multiple linear regression and artificial neural network. Chemometrics and Intelligent Laboratory Systems. 2015;143:7–15.

[226] Liu Y, Holder AJ. A quantum mechanical quantitative structure–property relationship study of the melting point of a variety of organosilicons. Journal of Molecular Graphics and Modelling. 2011;31:57–64.

[227] Liang G, Xu J, Liu L. QSPR analysis for melting point of fatty acids using genetic algorithm based multiple linear regression (GA-MLR). Fluid Phase Equilibria. 2013;353:15–21.

[228] Yan F, Xia S, Wang Q, Yang Z, Ma P. Predicting the melting points of ionic liquids by the Quantitative Structure Property Relationship method using a topological index. The Journal of Chemical Thermodynamics. 2013;62:196–200.

[229] Watkins M, Sizochenko N, Rasulev B, Leszczynski J. Estimation of melting points of large set of persistent organic pollutants utilizing QSPR approach. Journal of Molecular Modeling. 2016;22:55.

[230] Morrill JA, Byrd EF. Development of quantitative structure property relationships for predicting the melting point of energetic materials. Journal of Molecular Graphics and Modelling. 2015;62:190–201.

[231] Farahani N, Gharagheizi F, Mirkhani SA, Tumba K. Ionic liquids: Prediction of melting point by molecular-based model. Thermochimica Acta. 2012;549:17–34.

[232] Keshavarz MH, Pouretedal HR, Saberi E. A novel method for predicting melting point of ionic liquids. Process Safety and Environmental Protection. 2018;116:333–9.

[233] Aguirre CL, Cisternas LA, Valderrama JO. Melting-point estimation of ionic liquids by a group contribution method. International Journal of Thermophysics. 2012;33:34–46.

[234] Tarver CM, Chidester SK. On the Violence of High Explosive Reactions. Journal of Pressure Vessel Technology. 2005;127:39–48.

[235] Zeman S. Some predictions in the field of the physical thermal stability of nitramines. Thermochimica Acta. 1997;302:11–6.

[236] Zeman S. Sensitivities of High Energy Compounds. In: Klapötke TM, editor. High Energy Density Materials. Berlin Heidelberg: Springer; 2007. p. 195–271.

[237] Zeman S, Krupka M. New aspects of impact reactivity of polynitro compounds, Part II. Impact sensitivity as "the First Reaction" of polynitro arenes. Propellants, Explosives, Pyrotechnics. 2003;28:249–55.

[238] Zeman S, Krupka M. New aspects of impact reactivity of polynitro compounds, Part III. Impact sensitivity as a function of the imtermolecular interactions. Propellants, Explosives, Pyrotechnics. 2003;28:301–7.

[239] Zeman S. Study of the Initiation Reactivity of Energetic Materials. In: Armstrong R, editor. Energetics Science and Technology in Central Europe. Maryland: CALCE EPSC Press; 2012. p. 131–67.

[240] Goodarzi M, Chen T, Freitas MP. QSPR predictions of heat of fusion of organic compounds using Bayesian regularized artificial neural networks. Chemometrics and Intelligent Laboratory Systems. 2010;104:260–4.

[241] Atalar T, Zeman S. A new view of relationships of the N–N bond dissociation energies of cyclic nitramines. Part I. Relationships with heats of fusion. Journal of Energetic Materials. 2009;27:186–99.

[242] Chickos JS, Acree WE Jr, Liebman JF. Estimating solid–liquid phase change enthalpies and entropies. Journal of Physical and Chemical Reference Data. 1999;28:1535–673.

[243] Yu J, Sumathi R, Green WH. Accurate and efficient method for predicting thermochemistry of polycyclic aromatic hydrocarbons-bond-centered group additivity. Journal of the American Chemical Society. 2004;126:12685–700.

[244] Gharagheizi F, Salehi GR. Prediction of enthalpy of fusion of pure compounds using an artificial neural network-group contribution method. Thermochimica Acta. 2011;521:37–40.

[245] Mosaei Oskoei Y, Keshavarz MH. Improved method for reliable predicting enthalpy of fusion of energetic compounds. Fluid Phase Equilibria. 2012;326:1–14.

[246] Keshavarz MH. Prediction of enthalpy of fusion of non-aromatic energetic compounds containing nitramine, nitrate and nitro functional groups. Propellants, Explosives, Pyrotechnics. 2011;36:42–7.

[247] Keshavarz MH. A simple correlation for predicting heats of fusion of nitroaromatic carbocyclic energetic compounds. Journal of Hazardous Materials. 2008;150:387–93.

[248] Keshavarz MH. Predicting heats of fusion of nitramines. Indian Journal of Engineering and Materials Sciences. 2007;14:386–90.

[249] Semnani A, Keshavarz MH. Using molecular structure for reliable predicting enthalpy of melting of nitroaromatic energetic compounds. Journal of Hazardous Materials. 2010;178:264–72.

[250] Keshavarz MH. A new computer code for prediction of enthalpy of fusion and melting point of energetic materials. Propellants, Explosives, Pyrotechnics. 2015;40:150–5.

[251] Keshavarz MH, Akbarzadeh AR, Rahimi R, Jafari M, Pasandideh M, Sadeghi R. A reliable method for prediction of enthalpy of fusion in energetic materials using their molecular structures. Fluid Phase Equilibria. 2016;427:46–55.

[252] Ognichenko LN, Kuz'min VE, Gorb L, Muratov EN, Artemenko AG, Kovdienko NA, Polishchuk PG, Hill FC, Leszczynski J. New Advances in QSPR/QSAR Analysis of Nitrocompounds: Solubility, Lipophilicity, and Toxicity. In: Leszczynski J, Shukla MK, editors. Practical Aspects of Computational Chemistry II. Dordrecht, The Netherlands: Springer; 2012. p. 279–334.

[253] Poling BE, Prausnitz JM, Paul OCJ, Reid RC. The properties of gases and liquids. 5th ed. New York: McGraw-Hill; 2001.

[254] Jain A, Yang G, Yalkowsky SH. Estimation of melting points of organic compounds. Industrial & Engineering Chemistry Research. 2004;43:7618–21.

[255] Yalkowsky SH. Carnelley's rule and the prediction of melting point. Journal of Pharmaceutical Sciences. 2014;103:2629–34.

[256] Naef R, Acree WE. Calculation of five thermodynamic molecular descriptors by means of a general computer algorithm based on the group-additivity method: standard enthalpies of vaporization, sublimation and solvation, and entropy of fusion of ordinary organic molecules and total phase-change entropy of liquid crystals. Molecules. 2017;22:1059.

[257] Alnemrat S, Hooper JP. Predicting Temperature-Dependent Solid Vapor Pressures of Explosives and Related Compounds Using a Quantum Mechanical Continuum Solvation Model. The Journal of Physical Chemistry A. 2013;117:2035–43.

[258] Politzer P, Ma Y, Lane P, Concha MC. Computational prediction of standard gas, liquid, and solid-phase heats of formation and heats of vaporization and sublimation. International Journal of Quantum Chemistry. 2005;105:341–7.

[259] Keshavarz MH, Pouretedal HR. A new simple approach to predict entropy of fusion of nitroaromatic compounds. Fluid Phase Equilibria. 2010;298:24–32.

[260] Keshavarz MH, Zakinejad S, Esmailpour K. An improved simple method for prediction of entropy of fusion of energetic compounds. Fluid Phase Equilibria. 2013;340:52–62.

[261] Jain A, Yang G, Yalkowsky SH. Estimation of total entropy of melting of organic compounds. Industrial & Engineering Chemistry Research. 2004;43:4376–9.

[262] Dannenfelser R, Surendran N, Yalkowsky SH. Molecular symmetry and related properties. SAR and QSAR in Environmental Research. 1993;1:273–92.

[263] Dannenfelser R-M, Yalkowsky SH. Estimation of entropy of melting from molecular structure: A non-group contribution method. Industrial & Engineering Chemistry Research. 1996;35:1483–6.

[264] Yalkowsky SH, Alantary D. Estimation of melting points of organics. Journal of Pharmaceutical Sciences. 2018;107:1211–27.

[265] Foltz M. Aging of pentaerythritol tetranitrate (PETN), LLNL-TR-415057, Lawrence Livermore National Laboratory, Livermore; 2009.

[266] Bhattacharia SK, Maiti A, Gee RH, Weeks BL. Sublimation Properties of Pentaerythritol Tetranitrate Single Crystals Doped with Its Homologs. Propellants, Explosives, Pyrotechnics. 2012;37:563–8.

[267] Hikal WM, Weeks BL. Sublimation kinetics and diffusion coefficients of TNT, PETN, and RDX in air by thermogravimetry. Talanta. 2014;125:24–8.

[268] Bhattacharia S. Kinetics of Energetic Materials: Investigation of Sublimation and Decomposition. Chemistry Department, Texas Tech University; 2013. p. 120.

[269] Atkins PW, De Paula J. Physical chemistry. 9th ed. Oxford: Oxford University Press; 2010.

[270] Fried LE, Howard WM, Souers PC. CHEETAH 2.0 User's Manual (LLNL UCRL-MA-117541 Rev. 5), Lawrence Livermore National Laboratory, Livermore, CA; 1998.

[271] Mader CL. Numerical Modeling of Explosives and Propellants. 3th ed. Boca Raton: Taylor & Francis Group; 2008.

[272] Muthurajan H, Sivabalan R, Talawar MB, Asthana SN. Computer simulation for prediction of performance and thermodynamic parameters of high energy materials. Journal of Hazardous Materials. 2004;112:17–33.

[273] Curtiss LA, Redfern PC, Frurip DJ. Theoretical methods for computing enthalpies of formation of gaseous compounds. In: Lipkowitz KB, Boyd DB, editors. Reviews in Computational Chemistry. New York: Wiley; 2000. p. 147–211.

[274] Hu A, Larade B, Dudiy S, Abou-Rachid H, Lussier L-S, Guo H. Theoretical Prediction of Heats of Sublimation of Energetic Materials Using Pseudo-Atomic Orbital Density Functional Theory Calculations. Propellants, Explosives, Pyrotechnics. 2007;32:331–7.

[275] Ghosh MK, Cho SG, Choi CH. A priori prediction of heats of vaporization and sublimation by EFP2-MD. The Journal of Physical Chemistry B. 2014;118:4876–82.

[276] Salahinejad M, Le TC, Winkler DA. Capturing the crystal: prediction of enthalpy of sublimation, crystal lattice energy, and melting points of organic compounds. Journal of Chemical Information and Modeling. 2013;53:223–9.

[277] Zeman S, Krupka M. Some predictions of the heats of fusion, heats of sublimation and lattice energies of energetic materials. Chinese Journal of Energetic Materials (Hanneng Cailiao). 2002;10:27–33.

[278] Bondi AA. Physical Properties of Molecular Crystals Liquids, and Glasses. New York: John Wiley and Sons; 1968.

[279] Cundall RB, Palmer TF, Wood CE. Vapour pressure measurements on some organic high explosives. Journal of the Chemical Society, Faraday Transactions 1: Physical Chemistry in Condensed Phases. 1978;74:1339–45.

[280] Zeman S. New aspects of initiation reactivities of energetic materials demonstrated on nitramines. Journal of Hazardous Materials. 2006;132:155–64.

[281] Keshavarz MH, Shokrolahi A, Esmailpoor K, Zali A, Hafizi HR, Azamiamehraban J. Recent developments in predicting impact and shock sensitivities of energetic materials. Chinese Journal of Energetic Materials (HanNeng CaiLiao). 2008;16:113–20.

[282] Türker L. Recent Developments in the Theory of Explosive Materials. In: Janssen TJ, editor. Explosive Materials: Classification, Composition and Properties. New York: Nova Publisher; 2011. p. 1–52.

[283] Keshavarz MH. Important aspects of sensitivity of energetic compounds: a simple novel approach to predict electric spark sensitivity, Explosive materials: classification, composition and properties. New York: Nova Science Publishers; 2011. p. 103–23.

[284] Yan QL, Zeman S. Theoretical evaluation of sensitivity and thermal stability for high explosives based on quantum chemistry methods: a brief review. International Journal of Quantum Chemistry. 2013;113:1049–61.

[285] Zeman S, Jungová M. Sensitivity and performance of energetic materials. Propellants, Explosives, Pyrotechnics. 2016;41:426–51.

[286] Li G, Zhang C. Review of the molecular and crystal correlations on sensitivities of energetic materials. Journal of Hazardous Materials. 2020;398:122910.

[287] Politzer P, Murray JS. Impact sensitivity and the maximum heat of detonation. Journal of molecular modeling. 2015;21:1–11.

[288] Politzer P, Murray JS. Impact sensitivity and crystal lattice compressibility/free space. Journal of molecular modeling. 2014;20:1–8.

[289] Pospíšil M, Vávra P, Concha MC, Murray JS, Politzer P. Sensitivity and the available free space per molecule in the unit cell. Journal of Molecular Modeling. 2011;17:2569–74.

[290] Jungová M, Zeman S, Yan Q-L. Recent Advances in the Study of the Initiation of Nitramines by Impact Using Their 15N NMR Chemical Shifts. Central European Journal of Energetic Materials. 2014;11.

[291] Zhou Y, Du J-L, Long X-P, Shu Y-J. Impact sensitivity and nucleus-independent chemical shift for aromatic explosives. Molecular Simulation. 2013;39:716–20.

[292] Zohari N, Keshavarz MH, Seyedsadjadi SA. A link between impact sensitivity of energetic compounds and their activation energies of thermal decomposition. Journal of Thermal Analysis and Calorimetry. 2014;1–10.

[293] Keshavarz MH. A new general correlation for predicting impact sensitivity of energetic compounds. Propellants, Explosives, Pyrotechnics. 2013;38:754–60.

[294] Fang-qiang Y, Cong-zhong C, Shuai Z. Prediction of impact sensitivity of nitro energetic compounds by using structural parameters. Explosion and Shock Waves. 2013;1:012.

[295] Keshavarz MH, Motamedoshariati H, Moghayadnia R, Ghanbarzadeh M, Azarniamehraban J. Prediction of sensitivity of energetic compounds with a new computer code. Propellants, Explosives, Pyrotechnics. 2014;39:95–101.

[296] Kim H, Hwang S-n, Lee SK. QSPR analysis for the prediction of impact sensitivity of High Energy Density Materials (HEDM). Chinese Journal of Energetic Materials (HanNeng CaiLiao). 2012;275.

[297] Fayet G, Rotureau P. Development of simple QSPR models for the impact sensitivity of nitramines. Journal of Loss Prevention in the Process Industries. 2014;30:1–8.

[298] Prana V, Fayet G, Rotureau P, Adamo C. Development of validated QSPR models for impact sensitivity of nitroaliphatic compounds. Journal of Hazardous Materials. 2012;235:169–77.

[299] Fayet G, Rotureau P, Prana V, Adamo C. Global and local QSPR models to predict the impact sensitivity of nitro compounds. In: AIChE Spring Meeting 2012 & 8. Global Congress on Process Safety (GCPS). New York: AIChE; 2012.

[300] Fayet G, Rotureau P, Prana V, Adamo C. Global and local quantitative structure–property relationship models to predict the impact sensitivity of nitro compounds. Process Safety Progress. 2012;31:291–303.

[301] Xu J, Zhu L, Fang D, Wang L, Xiao S, Liu L, Xu W. QSPR studies of impact sensitivity of nitro energetic compounds using three-dimensional descriptors. Journal of Molecular Graphics and Modelling. 2012;36:10–9.

[302] Wang R, Wang YG, Liu H. Predicting Impact Sensitivity of Heterocyclic Nitroarenes from Molecular Structures Selected by Genetic Algorithm. In: Advanced Materials Research. Trans Tech Publ; 2012. p. 2550–3.

[303] Wang R, Jiang J, Pan Y. Prediction of impact sensitivity of nonheterocyclic nitroenergetic compounds using genetic algorithm and artificial neural network. Journal of Energetic Materials. 2012;30:135–55.

[304] Owens FJ, Jayasuriya K, Abrahmsen L, Politzer P. Computational analysis of some properties associated with the nitro groups in polynitroaromatic molecules. Chemical Physics Letters. 1985;116:434–8.

[305] Murray JS, Lane P, Politzer P, Bolduc PR. A relationship between impact sensitivity and the electrostatic potentials at the midpoints of C–NO2 bonds in nitroaromatics. Chemical Physics Letters. 1990;168:135–9.

[306] Vaullerin M, Espagnacq A, Morin-Allory L. Prediction of explosives impact sensitivity. Propellants, Explosives, Pyrotechnics. 1998;23:237–9.

[307] Fukuyama I, Ogawa T, Miyake A. Sensitivity and evaluation of explosive substances. Propellants, explosives, pyrotechnics. 1986;11:140–3.

[308] Brinck T, Murray JS, Politzer P. Quantitative determination of the total local polarity (charge separation) in molecules. Molecular Physics. 1992;76:609–17.

[309] Xiao H-M, Fan J-F, Gu Z-M, Dong H-S. Theoretical study on pyrolysis and sensitivity of energetic compounds:(3) Nitro derivatives of aminobenzenes. Chemical Physics. 1998;226:15–24.

[310] Politzer P, Murray JS, Markinas PL. Organic Energetic Compounds. New York: Nova Science; 1996.

[311] Rice BM, Hare JJ. A quantum mechanical investigation of the relation between impact sensitivity and the charge distribution in energetic molecules. The Journal of Physical Chemistry A. 2002;106:1770–83.

[312] Jane P, Murray PL. Effects of strongly electron-attracting components on molecular surface electrostatic potentials: application to predicting impact sensitivities of energetic molecules. Molecular Physics. 1998;93:187–94.

[313] Brill T, Oyumi Y. Thermal decomposition of energetic materials. 9. A relationship of molecular structure and vibrations to decomposition: polynitro-3, 3,7,7-tetrakis (trifluoromethyl)-2,4,6,8-tetraazabicyclo (3.3.0) octanes. The Journal of Physical Chemistry. 1986;90:2679–82.

[314] Edwards J, Eybl C, Johnson B. Correlation between sensitivity and approximated heats of detonation of several nitroamines using quantum mechanical methods. International Journal of Quantum Chemistry. 2004;100:713–9.

[315] Cao D-l, Shi W-j, Gao H-f. A theoretical prediction of the relationships between the impact sensitivity and electrostatic potential in strained cyclic explosive and application to H-bonded complex of nitrocyclohydrocarbon. Journal of Molecular Modeling. 2016;22:97.

[316] Oliveira MA, Borges I Jr. On the molecular origin of the sensitivity to impact of cyclic nitramines. International Journal of Quantum Chemistry. 2019;119:e25868.

[317] Zhang C, Shu Y, Huang Y, Zhao X, Dong H. Investigation of correlation between impact sensitivities and nitro group charges in nitro compounds. The Journal of Physical Chemistry B. 2005;109:8978–82.

[318] Bondarchuk SV. Quantification of impact sensitivity based on solid-state derived criteria. The Journal of Physical Chemistry A. 2018;122:5455–63.

[319] Cawkwell M, Manner V. Ranking the drop-weight impact sensitivity of common explosives using Arrhenius chemical rates computed from quantum molecular dynamics simulations. The Journal of Physical Chemistry A. 2019;124:74–81.

[320] Mathieu D. Sensitivity of energetic materials: Theoretical relationships to detonation performance and molecular structure. Industrial & Engineering Chemistry Research. 2017;56:8191–201.

[321] Cho S-G, No K-T, Goh E-M, Kim J-K, Shin J-H, Joo Y-D, Seong S-Y. Optimization of neural networks architecture for impact sensitivity of energetic molecules. Bulletin of the Korean Chemical Society. 2005;26:399–408.

[322] Keshavarz MH, Jaafari M. Investigation of the various structure parameters for predicting impact sensitivity of energetic molecules via artificial neural network. Propellants, Explosives, Pyrotechnics. 2006;31:216–25.

[323] Wang R, Jiang J, Pan Y, Cao H, Cui Y. Prediction of impact sensitivity of nitro energetic compounds by neural network based on electrotopological-state indices. Journal of Hazardous Materials. 2009;166:155–86.

[324] Kamlet MJ. The relationship of impact sensitivity with structure of organic high explosives. I. Polynitroaliphatic explosives. In: Sixth Symposium (International) on Detonation. Coronads, CA. 1976. p. 69–72.

[325] Kamlet MJ, Adolph HG. The relationship of impact sensitivity with structure of organic high explosives. II. Polynitroaromatic explosives. Propellants, Explosives, Pyrotechnics. 1979;4:30–4.

[326] Mullay J. A relationship between impact sensitivity and molecular electronegativity. Propellants, explosives, pyrotechnics. 1987;12:60–3.

[327] Mullay J. Relationships between impact sensitivity and molecular electronic structure. Propellants, explosives, pyrotechnics. 1987;12:121–4.

[328] McNesby K, Coffey C. Spectroscopic determination of impact sensitivities of explosives. The Journal of Physical Chemistry B. 1997;101:3097–104.

[329] Jain S. Energetics of propellants, fuels and explosives; a chemical valence approach. Propellants, explosives, pyrotechnics. 1987;12:188–95.

[330] Zeman S. New aspects of impact reactivity of polynitro compounds. Part IV. Allocation of polynitro compounds on the basis of their impact sensitivities. Propellants, Explosives, Pyrotechnics. 2003;28:308–13.

[331] Keshavarz MH, Pouretedal HR. Simple empirical method for prediction of impact sensitivity of selected class of explosives. Journal of Hazardous Materials. 2005;124:27–33.

[332] Keshavarz MH, Pouretedal HR, Semnani A. Novel correlation for predicting impact sensitivity of nitroheterocyclic energetic molecules. Journal of Hazardous Materials. 2007;141:803–7.

[333] Keshavarz MH. Prediction of impact sensitivity of nitroaliphatic, nitroaliphatic containing other functional groups and nitrate explosives. Journal of Hazardous Materials. 2007;148:648–52.

[334] Keshavarz MH, Zali A, Shokrolahi A. A simple approach for predicting impact sensitivity of polynitroheteroarenes. Journal of Hazardous Materials. 2009;166:1115–9.

[335] Keshavarz MH. Simple relationship for predicting impact sensitivity of nitroaromatics, nitramines, and nitroaliphatics. Propellants, Explosives, Pyrotechnics. 2010;35:175–81.

[336] Storm CB, Stine JR, Kramer JF. Sensitivity Relationships in Energetic Materials, in: Chemistry and Physics of Energetic Materials. Dordrecht: Springer; 1990. p. 605–39.

[337] Kamlet MJ, Adolph HG. Some comments regarding the sensitivities, thermal stabilities, and explosive performance characteristics of fluorodinitromethyl compounds. In: The Seventh Symp. (Int.) on Detonation. 1981. p. 60–7.

[338] Storm CB, Ryan RR, Ritchie JP, Hall JH, Bachrach SM. Structural basis of the impact sensitivities of 1-picryl-1,2,3-triazole, 2-picryl-1,2,3-triazole, 4-nitro-1-picryl-1,2,3-triazole, and 4-nitro-2-picryl-1,2,3-triazole. Journal of Physical Chemistry; (USA). 1989;93.

[339] Keshavarz MH, Esmaeilpour K, Khoshandam H, Keshavarz Z, Atabak HH, Damiri S, Afzali A. A novel method for the prediction of the impact sensitivity of quaternary ammonium-based energetic ionic liquids. Central European Journal of Energetic Materials. 2017;14.

[340] Hafner K, Klapötke TM, Schmid PC, Stierstorfer J. Synthesis and characterization of asymmetric 1,2-dihydroxy-5,5′-bitetrazole and selected nitrogen-rich derivatives. European Journal of Inorganic Chemistry. 2015;2015:2794–803.

[341] Zeman S, Valenta P, Zeman V, Jakubko J, Kamensky Z. Electric spark sensitivity of polynitro compounds: a comparison of some authors' results. Chinese Journal of Energetic Materials (Hanneng Cailiao). 1998;6:118–22.

[342] Friedldl Z, Kočí J. Electric spark sensitivity of nitramines. Part I. Aspects of molecular structure. Central European Journal of Energetic Materials. 2006;3:27–44.

[343] Zeman S, Pelikán W, Majzlík J, Kočí J. Electric Spark Sensitivity of Nitramines. Part II. A Problem of "Hot Spots". Central European Journal of Energetic Materials. 2006;3:45–51.

[344] Zeman V, Koci J, Zeman S. Electric spark sensitivity of polynitro compounds: Part II. A correlation with detonation velocities of some polynitro arenes. Chinese Journal of Energetic Materials (Hanneng Cailiao). 1999;7:127–32.

[345] Auzenau M, Roux M. Electric spark and ESD sensitivity of reactive solids. Part II: energy transfer mechanism and comprehensive study on E50, Propellants, Explosives. Pyrotechnics. 1995;20:96–101.

[346] Skinner D, Olson D, Block-Bolten A. Electrostatic discharge ignition of energetic materials. Propellants, Explosives, Pyrotechnics. 1998;23:34–42.

[347] Hosoya F, Shiino K, Itabashi K. Electric-spark sensitivity of Heat-Resistant Polynitroaromatic Compounds, Propellants, explosives. pyrotechnics. 1991;16:119–22.

[348] Zeman V, Koci J, Zeman S. Electric spark sensitivity of polynitro compounds: Part III. A correlation with detonation velocities of some nitramines. ENERGETIC MATERIALS-CHENGDU. 1999;7:172–5.

[349] Zeman S. The relationship between differential thermal analysis data and the detonation characteristics of polynitroaromatic compounds. Thermochimica Acta. 1980;41:199–212.

[350] Zeman S, Liu N. A new look on the electric spark sensitivity of nitramines. Defence Technology. 2020;16:10–7.

[351] Zeman S. Influence of the energy content and its outputs on sensitivity of polynitroarenes. Journal of Energetic Materials. 2019;37:445–58.

[352] Tan B, Li Z, Guo X, Li J, Han Y, Long X. Insight into electrostatic initiation of nitramine explosives. Journal of Molecular Modeling. 2017;23:1–9.

[353] Wang G, Xiao H, Ju X, Gong X. Calculation of detonation velocity, pressure, and electric sensitivity of nitro arenes based on quantum chemistry. Propellants, Explosives, Pyrotechnics. 2006;31:361–8.

[354] Wang GX, Xiao HM, Xu XJ, Ju XH. Detonation velocities and pressures, and their relationships with electric spark sensitivities for nitramines. Propellants, Explosives, Pyrotechnics. 2006;31:102–9.

[355] Keshavarz MH. Relationship between the electric spark sensitivity and detonation pressure. Indian Journal of Engineering and Materials Sciences. 2008;15:281–6.

[356] Keshavarz MH, Pouretedal HR, Semnani A. A simple way to predict electric spark sensitivity of nitramines. Indian Journal of Engineering and Materials Sciences. 2008;15:505–9.

[357] Keshavarz MH, Pouretedal HR, Semnani A. Simple way to predict electrostatic sensitivity of nitroaromatic compounds. Khimiya. 2008;17:470–84.

[358] Keshavarz MH, Pouretedal HR, Semnani A. Reliable prediction of electric spark sensitivity of nitramines: A general correlation with detonation pressure. Journal of Hazardous Materials. 2009;167:461–6.

[359] Keshavarz MH, Keshavarz Z. Relation between electric spark sensitivity and impact sensitivity of nitroaromatic energetic compounds. Zeitschrift für anorganische und allgemeine Chemie. 2016;642:335–42.

[360] Türker L. Contemplation on spark sensitivity of certain nitramine type explosives. Journal of Hazardous Materials. 2009;169:454–9.

[361] Zhi C, Cheng X, Zhao F. The correlation between electric spark sensitivity of polynitroaromatic compounds and their molecular electronic properties. Propellants, Explosives, Pyrotechnics. 2010;35:555–60.

[362] Keshavarz MH. Theoretical prediction of electric spark sensitivity of nitroaromatic energetic compounds based on molecular structure. Journal of Hazardous Materials. 2008;153:201–6.

[363] Keshavarz MH, Moghadas MH, Kavosh Tehrani M. Relationship between the electrostatic sensitivity of nitramines and their molecular structure. Propellants, Explosives, Pyrotechnics. 2009;34:136–41.

[364] Zeman S, Koci J. Electric spark sensitivity of polynitro compounds: part IV. A relation to thermal decomposition parameters. Chinese Journal of Energetic Materials (Hanneng Cailiao). 2000;8:18–26.

[365] Keshavarz MH. A novel approach for the prediction of electric spark sensitivity of polynitroarenes based on the measured data from a new instrument. Central European Journal of Energetic Materials. 2019;16:65–76.

[366] Zeman S, Majzlík J. Electric spark sensitivity of polynitro arenes Part I. A comparison of two instruments. Central European Journal of Energetic Materials. 2007;4:15–24.

[367] Keshavarz MH, Damiri S, Bagheri V. Recent advances for prediction of electric spark and shock sensitivities of organic compounds containing energetic functional groups to assess reliable models. Process Safety and Environmental Protection. 2019;131:9–15.

[368] Keshavarz MH. Two novel correlations for prediction of electric spark sensitivity of nitramines based on the experimental data of the new instrument. Zeitschrift für anorganische und allgemeine Chemie. 2018;644:1607–10.

[369] Zeman S, Pelikán V, Majzlík J, Friedl Z, Kočí J. Electric spark sensitivity of nitramines. Part I. Aspects of molecular structure. Central European Journal of Energetic Materials. 2006;3:27–44.

[370] Nazari B, Keshavarz MH, Jafari M, Jafari F. A novel approach for prediction of sensitivity toward the electrical discharge of quaternary ammonium-based energetic ionic liquids or salts. Zeitschrift für anorganische und allgemeine Chemie. 2018;644:1153–7.

[371] Dippold A. Nitrogen-rich energetic materials based on 1,2,4-triazole derivatives. LMU; 2013.

[372] Price D. Examination of some proposed relations among HE sensitivity data. Journal of Energetic Materials. 1985;3:239–54.

[373] Tan B, Long X, Peng R, Li H, Jin B, Chu S, Dong H. Two important factors influencing shock sensitivity of nitro compounds: bond dissociation energy of X–NO 2 (X=C, N, O) and Mulliken charges of nitro group. Journal of Hazardous Materials. 2010;183:908–12.

[374] Keshavarz MH, Motamedoshariati H, Pouretedal HR, Tehrani MK, Semnani A. Prediction of shock sensitivity of explosives based on small-scale gap test. Journal of Hazardous Materials. 2007;145:109–12.

[375] Dobratz B. LLNL Explosives Handbook: Properties of Chemical Explosives and Explosives and Explosive Simulants. Lawrence Livermore National Lab., CA (USA); 1981.

[376] Keshavarz MH, Pouretedal HR, Tehrani MK, Semnani A. Predicting shock sensitivity of energetic compounds. Asian Journal of Chemistry. 2008;20:1025.

[377] Dobratz BM, Crawford PC, Handbook LE. Properties of Chemical Explosives and Explosives Simulants. Lawrence Livermore National Lab., CA (USA); 1985.

[378] Cooper PW. Explosives Engineering. John Wiley & Sons; 1996.

[379] Kobylkin IF. Calculation of the critical detonation diameter of explosive charges using data on their shock-wave initiation. Combustion, Explosion and Shock Waves. 2006;42:223–6.

[380] Pepekin VI, Gubina TV. On the critical diameter and detonability of explosives. Russian Journal of Physical Chemistry B, Focus on Physics. 2011;5:813–5.

[381] Kobylkin IF. Critical detonation diameter of highly desensitized low-sensitivity explosive formulations. Combustion, Explosion, and Shock Waves. 2009;45:732.

[382] Kobylkin IF. Critical detonation diameter of industrial explosive charges: Effect of the casing. Combustion, Explosion, and Shock Waves. 2011;47:96.

[383] Keshavarz MH, Klapötke TM. A novel method for prediction of the critical diameter of solid pure and composite high explosives to assess their explosion safety in an industrial setting. Journal of Energetic Materials. 2019;37:331–9.

[384] Matyáš R, Šelešovský J, Musil T. Sensitivity to friction for primary explosives. Journal of Hazardous Materials. 2012;213:236–41.

[385] Jungová M, Zeman S, Husarová A. Friction sensitivity of nitramines. Part II. Comparison with thermal reactivity. Chinese Journal of Energetic Materials (HanNeng CaiLiao). 2011;19:607–9.

[386] Friedl Z, Jungová M, Zeman S, Husarová A. Friction sensitivity of nitramines. Part IV: Links to surface electrostatic potentials. Chinese Journal of Energetic Materials (HanNeng CaiLiao). 2012;19:613–5.

[387] Jungová M, Zeman S, Husarová A. Friction Sensitivity of Nitramines. Part I: Comparison with Impact Sensitivity and Heat of Fusion. Chinese Journal of Energetic Materials (HanNeng CaiLiao). 2012;19:603–6.

[388] Zeman S, Jungová M, Husarová A. Friction sensitivity of nitramines. Part III: Comparison with detonation performance. Chinese Journal of Energetic Materials (Hanneng Cailiao). 2011;19:610–2.

[389] Le Roux JJ. The dependence of friction sensitivity of primary explosives upon rubbing surface roughness. Propellants, Explosives, Pyrotechnics. 1990;15:243–7.

[390] Keshavarz MH, Hayati M, Ghariban-Lavasani S, Zohari N. A new method for predicting the friction sensitivity of nitramines. Central European Journal of Energetic Materials. 2015;12:215–27.

[391] Klapötke TM. New nitrogen-rich high explosives. In: High Energy Density Materials. Springer; 2007. p. 85–121.

[392] Jafari M, Keshavarz MH, Joudaki F, Mousaviazar A. A simple method for predicting friction sensitivity of quaternary ammonium-based energetic ionic liquids. Propellants, Explosives, Pyrotechnics. 2018;43:568–73.

[393] Dippold AA, Klapötke TM. A study of dinitro-bis-1,2,4-triazole-1, 1′-diol and derivatives: design of high-performance insensitive energetic materials by the introduction of N-oxides. Journal of the American Chemical Society. 2013;135:9931–8.

[394] Pourmortazavi SM, Rahimi-Nasrabadi M, Kohsari I, Hajimirsadeghi SS. Non-isothermal kinetic studies on thermal decomposition of energetic materials: KNF and NTO. Journal of Thermal Analysis and Calorimetry. 2011;110:857–63.

[395] Cusu JP, Musuc AM, Matache M, Oancea D. Kinetics of exothermal decomposition of some ketone-2,4-dinitrophenylhydrazones. Journal of Thermal Analysis and Calorimetry. 2012;110:1259–66.

[396] Lee J-S, Hsu C-K, Chang C-L. A study on the thermal decomposition behaviors of PETN, RDX, HNS and HMX. Thermochimica Acta. 2002;392:173–6.

[397] Chen ZX, Xiao H. Impact sensitivity and activation energy of pyrolysis for tetrazole compounds. International Journal of Quantum Chemistry. 2000;79:350–7.

[398] Sinditskii VP, Smirnov SP, Egorshev VY. Thermal decomposition of NTO: an explanation of the high activation energy. Propellants, Explosives, Pyrotechnics. 2007;32:277.

[399] Zeman S, Dimun M, Truchlik Š. The relationship between kinetic data of the low-temperature thermolysis and the heats of explosion of organic polynitro compounds. Thermochimica Acta. 1984;78:181–209.

[400] Zeman S. Thermal stabilities of polynitroaromatic compounds and their derivatives. Thermochimica Acta. 1979;31:269–83.

[401] Zeman S. Non-isothermal differential thermal analysis in the study of the initial state of the thermal decomposition of polynitroaromatic compounds in the condensed state. Thermochimica Acta. 1980;39:117–24.

[402] Zeman S. The thermoanalytical study of some aminoderivatives of 1,3,5-trinitrobenzene. Thermochimica acta. 1993;216:157–68.

[403] Zeman S. Relationship between the Arrhenius Parameters of the Low-temperature Thermolysis and the 13C and 15N Chemical Shifts of Nitramines. Thermochimica acta. 1992;202:191–200.

[404] Zeman S. Kinetic compensation effect and thermolysis mechanisms of organic polynitroso and polynitro compounds. Thermochimica Acta. 1997;290:199–217.

[405] Fathollahi M, Sajady H. QSPR modeling of decomposition temperature of energetic cocrystals using artificial neural network. Journal of Thermal Analysis and Calorimetry. 2018;133:1663–72.

[406] Dong J, Yan Q-L, Liu P-J, He W, Qi X-F, Zeman S. The correlations among detonation velocity, heat of combustion, thermal stability and decomposition kinetics of nitric esters. Journal of Thermal Analysis and Calorimetry. 2018;131:1391–403.

[407] Zeman S. Analysis and prediction of the Arrhenius parameters of low-temperature thermolysis of nitramines by means of the 15N NMR spectroscopy. Thermochimica Acta. 1999;333:121–9.

[408] Zeman S, Jalový Z. Heats of fusion of polynitro derivatives of polyazaisowurtzitane. Thermochimica Acta. 2000;345:31–8.

[409] Zeman S, Friedl Z. Relationship between electronic charges at nitrogen atoms of nitro groups and thermal reactivity of nitramines. Journal of Thermal Analysis and Calorimetry. 2004;77:217–24.

[410] Sorescu DC, Rice BM, Thompson DL. Molecular packing and molecular dynamics study of the transferability of a generalized nitramine intermolecular potential to non-nitramine crystals. The Journal of Physical Chemistry A. 1999;103:989–98.

[411] Kissinger HE. Reaction kinetics in differential thermal analysis. Analytical Chemistry. 1957;29:1702–6.

[412] Zeman S. Modified Evans–Polanyi–Semenov relationship in the study of chemical micromechanism governing detonation initiation of individual energetic materials. Thermochimica Acta. 2002;384:137–54.

[413] Keshavarz MH. A new method to predict activation energies of nitroparaffins. Indian Journal of Engineering and Materials Sciences. 2009;16:429–32.

[414] Manelis GB. Problemy Kinetiki Elementarnykh Khimicheskikh Reaktsii (Problems of the Kinetics of Primary Chemical Reactions). Moscow: Nauka; 1973.

[415] Keshavarz MH. Simple method for prediction of activation energies of the thermal decomposition of nitramines. Journal of Hazardous Materials. 2009;162:1557–62.

[416] Keshavarz MH, Pouretedal HR, Shokrolahi A, Zali A, Semnani A. Predicting activation energy of thermolysis of polynitro arenes through molecular structure. Journal of Hazardous Materials. 2008;160:142–7.
[417] Keshavarz MH, Zohari N, Seyedsadjadi SA. Validation of improved simple method for prediction of activation energy of the thermal decomposition of energetic compounds. Journal of Thermal Analysis and Calorimetry. 2013;114:497–510.
[418] Talawar MB, Sivabalan R, Mukundan T, Muthurajan H, Sikder AK, Gandhe BR, Rao AS. Environmentally compatible next generation green energetic materials (GEMs). Journal of Hazardous materials. 2009;161:589–607.
[419] Klapötke TM. Chemistry of High-Energy Materials. 2th ed. De Gruyter; 2012.
[420] Ando T, Fujimoto Y, Morisaki S. Analysis of differential scanning calorimetric data for reactive chemicals. Journal of Hazardous Materials. 1991;28:251–80.
[421] Saraf SR, Rogers WJ, Mannan MS. Prediction of reactive hazards based on molecular structure. Journal of Hazardous Materials. 2003;98:15–29.
[422] Fayet G, Joubert L, Rotureau P, Adamo C. On the use of descriptors arising from the conceptual density functional theory for the prediction of chemicals explosibility. Chemical Physics Letters. 2009;467:407–11.
[423] Fayet G, Rotureau P, Joubert L, Adamo C. On the prediction of thermal stability of nitroaromatic compounds using quantum chemical calculations. Journal of hazardous materials. 2009;171:845–50.
[424] Fayet G, Rotureau P, Joubert L, Adamo C. QSPR modeling of thermal stability of nitroaromatic compounds: DFT vs. AM1 calculated descriptors. Journal of molecular modeling. 2010;16:805–12.
[425] Fayet G, Del Rio A, Rotureau P, Joubert L, Adamo C. Predicting the thermal stability of nitroaromatic compounds using chemoinformatic tools. Molecular Informatics. 2011;30:623–34.
[426] Fayet G, Rotureau P, Joubert L, Adamo C. Development of a QSPR model for predicting thermal stabilities of nitroaromatic compounds taking into account their decomposition mechanisms. Journal of Molecular Modeling. 2011;17:2443–53.
[427] Fayet G, Rotureau P, Adamo C. On the development of QSPR models for regulatory frameworks: The heat of decomposition of nitroaromatics as a test case. Journal of Loss Prevention in the Process Industries. 2013;26:1100–5.
[428] Keshavarz MH, Ghani K, Asgari A. A new method for predicting heats of decomposition of nitroaromatics. Zeitschrift für anorganische und allgemeine Chemie. 2015;641:1818–23.
[429] Lu Y, Ng D, Mannan MS. Prediction of the reactivity hazards for organic peroxides using the QSPR approach. Industrial & Engineering Chemistry Research. 2010;50:1515–22.
[430] Prana V, Rotureau P, Fayet G, André D, Hub S, Vicot P, Rao L, Adamo C. Prediction of the thermal decomposition of organic peroxides by validated QSPR models. Journal of Hazardous Materials. 2014;276:216–24.
[431] Pan Y, Zhang Y, Jiang J, Ding L. Prediction of the self-accelerating decomposition temperature of organic peroxides using the quantitative structure–property relationship (QSPR) approach. Journal of Loss Prevention in the Process Industries. 2014;31:41–9.
[432] Gao Y, Xue Y, Lü Z-g, Wang Z, Chen Q, Shi N, Sun F. Self-accelerating decomposition temperature and quantitative structure–property relationship of organic peroxides. Process Safety and Environmental Protection. 2015;94:322–8.
[433] Zohari N, Keshavarz MH, Dalaei Z. Prediction of decomposition onset temperature and heat of decomposition of organic peroxides using simple approaches. Journal of Thermal Analysis and Calorimetry. 2016;125:887–96.
[434] Wakakura M, Iiduka Y. Trends in chemical hazards in Japan. Journal of Loss prevention in the process industries. 1999;12:79–84.

[435] Keshavarz MH, Pouretedal HR, Semnani A. Relationship between thermal stability and molecular structure of polynitro arenes. Indian Journal of Engineering and Materials Sciences. 2009;16:61–4.

[436] Keshavarz MH, Mousaviazar A, Hayaty M. A novel approach for assessment of thermal stability of organic azides through prediction of their temperature of maximum mass loss. Journal of Thermal Analysis and Calorimetry. 2017;129:1659–65.

[437] Kumari D, Anjitha SG, Pant CS, Patil M, Singh H, Banerjee S. Synthetic approach to novel azido esters and their utility as energetic plasticizers. RSC Advances. 2014;4:39924–33.

[438] Fedoroff BT, Sheffield OE, Reese EF, Sheffield OE, Clift GD, Dunkle CG, Walter H, Mclean DC. In: Encyclopedia of Explosives and Related Items, Part 2700, Picatinny Arsenal. Dower, NJ. 1960.

[439] Keshavarz MH, Moradi S, Saatluo BE, Rahimi H, Madram AR. A simple accurate model for prediction of deflagration temperature of energetic compounds. Journal of Thermal Analysis and Calorimetry. 2013;112:1453–63.

[440] Drees D, Löffel D, Messmer A, Schmid K. Synthesis and characterization of azido plasticizer. Propellants, Explosives, Pyrotechnics. 1999;24:159–62.

[441] Keshavarz MH, Nazari B, Jafari M, Yazdani Z. A novel and simple approach for predicting activation energy of thermolysis of some selected ionic liquids. Journal of Thermal Analysis and Calorimetry. 2018;134:2383–90.

[442] Cao Y, Mu T. Comprehensive investigation on the thermal stability of 66 ionic liquids by thermogravimetric analysis. Industrial & Engineering Chemistry Research. 2014;53:8651–64.

[443] Keshavarz MH, Pouretedal HR, Saberi E. A new method for predicting decomposition temperature of imidazolium-based energetic ionic liquids. Zeitschrift für anorganische und allgemeine Chemie. 2017;643:171–9.

[444] Schneider S, Hawkins T, Rosander M, Mills J, Vaghjiani G, Chambreau S. Liquid azide salts and their reactions with common oxidizers IRFNA and N 2 O 4. Inorganic Chemistry. 2008;47:6082–9.

[445] Zohari N, Abrishami F, Zeynali V. Prediction of decomposition temperature of azole-based energetic compounds in order to assess of their thermal stability. Journal of Thermal Analysis and Calorimetry. 2019;141:1453–63.

[446] Kumar D, Mitchell LA, Parrish DA, Jean'Ne MS. Asymmetric N, N′-ethylene-bridged azole-based compounds: Two way control of the energetic properties of compounds. Journal of Materials Chemistry A. 2016;4:9931–40.

[447] Koci Jí, Zeman V, Zeman S. Electric spark sensitivity of polynitro compounds. Part V. A relationship between electric spark and impact sensitivities of energetic materials. Chinese Journal of Energetic Materials (HanNeng CaiLiao). 2001;9:60–5.

[448] Zohari N, Keshavarz MH, Seyedsadjadi SA. A novel method for risk assessment of electrostatic sensitivity of nitroaromatics through their activation energies of thermal decomposition. Journal of Thermal Analysis and Calorimetry. 2014;115:93–100.

[449] Keshavarz MH, Hayati M, Ghariban-Lavasani S, Zohari N. Relationship between activation energy of thermolysis and friction sensitivity of cyclic and acyclic nitramines. Zeitschrift für anorganische und allgemeine Chemie. 2016;642:182–8.

[450] Keshavarz MH, Zohari N, Seyedsadjadi SA. Relationship between electric spark sensitivity and activation energy of the thermal decomposition of nitramines for safety measures in industrial processes. Journal of Loss Prevention in the Process Industries. 2013;26:1452–6.

[451] Zohari N, Seyed-Sadjadi SA, Marashi-Manesh S. The Relationship between Impact Sensitivity of Nitroaromatic Energetic Compounds and their Electrostatic Sensitivity. Central European Journal of Energetic Materials. 2016;13:427–43.

[452] Wu Y-Q, Huang F-L. A microscopic model for predicting hot-spot ignition of granular energetic crystals in response to drop-weight impacts. Mechanics of Materials. 2011;43:835–52.

[453] Keshavarz MH, Ghaffarzadeh M, Omidkhah MR, Farhadi K. New correlation between electric spark and impact sensitivities of nitramine energetic compounds for assessment of their safety. Zeitschrift für anorganische und allgemeine Chemie. 2017;643:1227–31.

[454] Ferdowsi M, Yazdani F, Omidkhah MR, Keshavarz MH. A general relationship between electric spark and impact sensitivities of nitroaromatics and nitramines. Zeitschrift für anorganische und allgemeine Chemie. 2018;644:1623–8.

[455] Östmark H, Bergman H, Ekvall K, Langlet A. A study of the sensitivity and decomposition of 1,3,5-trinitro-2-oxo-1,3,5-triazacyclo-hexane. Thermochimica Acta. 1995;260:201–16.

[456] Keshavarz MH, Ghaffarzadeh M, Omidkhah MR, Farhadi K. Correlation between Shock Sensitivity of Nitramine Energetic Compounds based on Small-scale Gap Test and Their Electric Spark Sensitivity. Zeitschrift für anorganische und allgemeine Chemie. 2017;643:2158–62.

[457] Tan B, Long X, Peng R, Li H, Jin B, Chu S. On the shock sensitivity of explosive compounds with small-scale gap test. The Journal of Physical Chemistry A. 2011;115:10610–6.

[458] Ferdowsi M, Yazdani F, Omidkhah MR, Keshavarz MH. Reliable prediction of shock sensitivity of energetic compounds based on small-scale gap test through their electric spark sensitivity. Zeitschrift für anorganische und allgemeine Chemie. 2018;644:888–92.

[459] Hobbs ML, Baer MR. Calibrating the BKW-EOS with a large product species data base and measured CJ properties. In: Proc. of the 10th Symp. (International) on Detonation, ONR. 1993. p. 409.

[460] Pedley JB. Thermochemical Data of Organic Compounds. Springer; 2012.

[461] Vadhe PP, Pawar RB, Sinha RK, Asthana SN, Rao AS. Cast aluminized explosives (review). Combustion, Explosion, and Shock Waves. 2008;44:461–77.

[462] Manaa MR, Fried LE, Kuo I-FW. Determination of enthalpies of formation of energetic molecules with composite quantum chemical methods. Chemical Physics Letters. 2016;648:31–5.

[463] Frisch MJ, Trucks GW, Schlegel HB, Scuseria GE, Robb MA, Cheeseman JR, Scalmani G, Barone V, Mennucci B, Petersson GA, Nakatsuji H, Caricato M, Li X, Hratchian HP, Izmaylov AF, Bloino J, Zheng G, Sonnenberg JL, Hada M, Ehara M, Toyota K, Fukuda R, Hasegawa J, Ishida M, Nakajima T, Honda Y, Kitao O, Nakai H, Vreven T, Montgomery JA, Peralta JE, Ogliaro F, Bearpark M, Heyd JJ, Brothers E, Kudin KN, Staroverov VN, Kobayashi R, Normand J, Raghavachari K, Rendell A, Burant JC, Iyengar SS, Tomasi J, Cossi M, Rega N, Millam JM, Klene M, Knox JE, Cross JB, Bakken V, Adamo C, Jaramillo J, Gomperts R, Stratmann RE, Yazyev O, Austin AJ, Cammi R, Pomelli C, Ochterski JW, Martin RL, Morokuma K, Zakrzewski VG, Voth GA, Salvador P, Dannenberg JJ, Dapprich S, Daniels AD, Farkas O, Foresman JB, Ortiz JV, Cioslowski J, Fox DJ. GAUSSIAN 09 (Revision A. 1). Wallingford: Gaussian, Inc.; 2009.

[464] Haynes WM. CRC Handbook of Chemistry and Physics. CRC press; 2014.

[465] Stevens WR, Ruscic B, Baer T. Heats of formation of C6H5•, C6H5+, and C6H5NO by threshold photoelectron photoion coincidence and active thermochemical tables analysis. The Journal of Physical Chemistry A. 2010;114:13134–45.

[466] Simmie JM. A database of formation enthalpies of nitrogen species by compound methods (CBS-QB3, CBS-APNO, G3, G4). The Journal of Physical Chemistry A. 2015;119:10511–26.

[467] Byrd EF, Rice BM. A comparison of methods to predict solid phase heats of formation of molecular energetic salts. The Journal of Physical Chemistry A. 2009;113:345–52.

[468] Verevkin SP, Emel'yanenko VN, Zaitsau DH, Heintz A, Muzny CD, Frenkel M. Thermochemistry of imidazolium-based ionic liquids: experiment and first-principles calculations. Physical Chemistry Chemical Physics. 2010;12:14994–5000.

[469] Emel'yanenko VN, Verevkin SP, Heintz A. The gaseous enthalpy of formation of the ionic liquid 1-butyl-3-methylimidazolium dicyanamide from combustion calorimetry, vapor pressure measurements, and ab initio calculations. Journal of the American Chemical Society. 2007;129:3930–7.

[470] Zhang Z-H, Tan Z-C, Sun L-X, Jia-Zhen Y, Lv X-C, Shi Q. Thermodynamic investigation of room temperature ionic liquid: The heat capacity and standard enthalpy of formation of EMIES. Thermochimica Acta. 2006;447:141–6.

[471] Ye C, Xiao J-C, Twamley B, Shreeve JnM. Energetic salts of azotetrazolate, iminobis(5-tetrazolate) and 5,5'-bis(tetrazolate). Chemical communications. 2005;2750–2.

[472] Gutowski KE, Rogers RD, Dixon DA. Accurate thermochemical properties for energetic materials applications. II. Heats of formation of imidazolium-, 1,2,4-triazolium-, and tetrazolium-based energetic salts from isodesmic and lattice energy calculations. The Journal of Physical Chemistry B. 2007;111:4788–800.

[473] Pedley JB, Naylor RD, Kirby SP. Thermochemical Data of Organic Compounds. 2nd ed. New York: Chapman and Hall Ltd.; 1986.

[474] Darwich C, Klapötke TM, Welch JM, Suceska M. Synthesis and characterization of 3,4,5-triamino-1,2,4-triazolium 5-nitrotetrazolate. Propellants, Explosives, Pyrotechnics. 2007;32:235–43.

[475] Darwich C, Klapötke TM, Sabaté CM. 1,2,4-Triazolium-cation-based energetic salts. Chemistry–A European Journal. 2008;14:5756–71.

[476] Xue H, Shreeve JM. Energetic Ionic Liquids from Azido Derivatives of 1,2,4-Triazole. Advanced Materials. 2005;17:2142–6.

[477] Xue H, Twamley B, Jean'ne MS. Energetic salts of substituted 1,2,4-triazolium and tetrazolium 3,5-dinitro-1,2,4-triazolates. Journal of Materials Chemistry. 2005;15:3459–65.

Index

www.ingramcontent.com/pod-product-compliance
Lightning Source LLC
Chambersburg PA
CBHW082109220326
41598CB00066BA/5932